LONGMAN SCIENCE

PHYSICAL

Roland Scal

Longman Science: Physical

Pearson Education, 10 Bank Street, White Plains, NY 10606 USA

Staff credits: The people who made up the ***Longman Science: Physical*** team, representing editorial, production, design, manufacturing, and marketing are John Ade, Margaret Antonini, Rhea Banker, Gina DiLillo, Ed Lamprich, Tara Maceyak, Liza Pleva, Barbara Sabella, Tania Saiz-Sousa, Susan Saslow, and Patricia Wosczyk.
Text design and composition: The Quarasan Group, Inc.
Text font: 12.5/16 Minion Regular
Photo and Illustration credits: See page 155.

Library of Congress Cataloging-in-Publication Data
Scal, Roland.
Longman Science: Physical / Contributor: Scal, Roland.
p. cm.
Includes index.
ISBN-13: 978-0-13-267941-1
ISBN-10: 0-13-267941-8
1. Science—Study and teaching (Secondary)
I. Pearson Education, Inc.
Q181.L824 2011
500—dc22

ISBN-13: 978-0-13-267941-1
ISBN-10: 0-13-267941-8

Printed in the United States of America
1 2 3 4 5 6 7 8 9 10—V011—15 14 13 12 11

Contents

Getting Started

Introduction

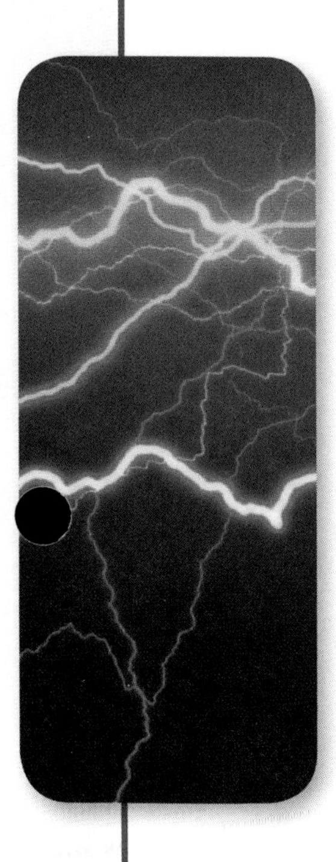

Part 1

Matter

Part 2

Sound and Light

Getting Started Introduction

What Is Science?

Science is the study of the natural world. It is the study of Earth and other planets. It is the study of the animals and plants on Earth. Science is the study of every living and nonliving thing around us.

Scientists study our world. They ask questions and look for answers. Scientists watch carefully. Then they try to understand.

Scientists study Earth and other planets. ▼

▲ Scientists study animals, like this koala. They also study plants, like this tree.

Scientists study different kinds of matter and energy. ▶

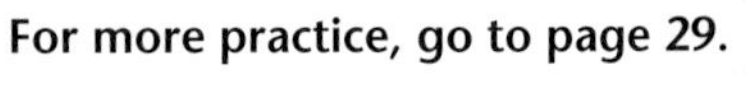

For more practice, go to page 29.

The Sciences

Scientists study different kinds of sciences. In this book you'll learn about physical science, Earth science, and life science.

Physical Science

Physical science is the study of matter. Air is matter. Rocks are matter. And water is matter. They are nonliving matter. Physical scientists study mostly nonliving matter.

Physical science is also the study of energy. Electricity is an example of energy. Sound and light are also **types** of energy.

---

types: kinds

▲ Electricity is a type of energy. Lightning is electricity.

◀ The boy in this photograph is living matter. Everything else is nonliving matter.

Earth Science

Earth science is the study of the Earth and outer space. It is the study of the land, water, and air on the Earth. It is the study of all the materials that make up the Earth and everything else in the universe. Earth science is the study of the sun you see in the day. It is the study of the stars you see at night. It is also the study of other planets far away from Earth.

▲ Hot liquid rock, called lava, comes out of volcanoes when they erupt.

▲ Water, air, land, and living things make up Earth's environment.

This is the planet Saturn. It is far away from Earth. ▼

Life Science

Life science is the study of living things on the Earth. Animals are living things. Plants are also living things.

▲ Some scientists study only frogs.

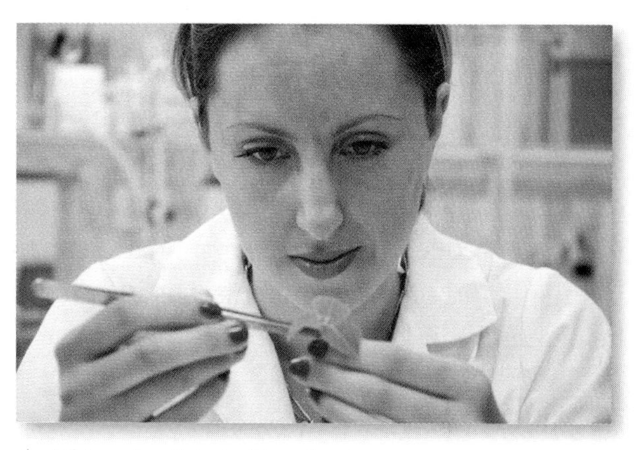

▲ This scientist studies plants.

▲ Some scientists study only plants and animals that live underwater.

Before You Go On

1. What is the environment?
2. What are some examples of matter?
3. What are some examples of energy?

For more practice, go to pages 30–32.

Meet a Scientist

Albert Einstein was a scientist and a mathematician. He was born in Germany in 1879. As a child, he was slow to begin to speak, and once in school, he had difficulties with many of his teachers. But Einstein is now known to have been one of the most **brilliant** human beings who ever lived.

Einstein enjoyed using his imagination, and he believed that imagination was far more important than knowledge. He used his imagination to think out math and science problems. His imagination helped him think up many of his theories.

---

brilliant: smart, intelligent

Portrait of physicist Albert Einstein, probably taken in 1931, by an unknown photographer ▶

▲ The formula for Einstein's theory of relativity

Einstein is perhaps most famous for his theory of relativity. It is an extremely complex theory dealing with how an object's speed changes depending on where you are in relation to the object. For example, if you look at an airplane high in the sky from the ground, the plane seems to go very slowly. But if the plane is right above you, it seems to go much faster. And if you are inside the plane, it doesn't seem to you to be moving at all.

Einstein won a Nobel Prize in 1922 for his work in theoretical physics and for his theory about the photoelectric effect, which led to the invention of the television. In his later years, Einstein became a world peace advocate.

▲ Washington, D.C., Memorial to Albert Einstein

◀ Israeli 5 Lirot bill with portrait of Einstein

The Scientific Method

How do scientists find out information about the world? How do they understand the information? How do they show others their ideas are correct?

Scientists use the scientific method. This is how scientists find out about the world. Five important steps of the scientific method are (1) asking questions, (2) making a **hypothesis**, (3) testing the hypothesis, (4) observing, and (5) **drawing conclusions**.

---

hypothesis: a guess; an idea
drawing conclusions: deciding about something

◀ Scientists look at things carefully.

Scientists measure things. This scientist is using beakers to make measurements. ▼

Asking Questions

Scientists ask questions about the world. They begin with things they know. They ask questions about things they don't know.

For example, you have questions about plants. You know seeds grow into plants. But you don't know how. You ask, "What is something that seeds need to grow?"

▼ This plant grew from a seed.

▲ Scientists look at the world and ask questions.

▼ These are seeds.

Before You Go On

1. How do scientists show others their ideas are correct?
2. What are the five steps of the scientific method?
3. What's an example of asking questions?

Making a Hypothesis

After asking a question, scientists try to guess the answer. This is called making a hypothesis.

You asked, "What is something that seeds need to grow?" You think about what you already know about plants. Then you make a guess. You think, "The seeds need water to grow." This is your hypothesis.

▲ You get a paper towel, a glass jar, and some beans.

▲ You put a bean, paper towel, and water in the jar. You put a bean and dry paper towel in another jar.

Testing the Hypothesis

After scientists make a hypothesis, they test it. They find out if their idea is correct. Doing an experiment is a good way to test a hypothesis.

To test your hypothesis about seeds, you think of an experiment. You know beans are big seeds. You have some beans in the kitchen. You put paper towels and beans into two jars. Then you add water to one of the jars.

Observing

Scientists observe, or look, listen, touch, and think. Observing is an important part of the scientific method.

During your bean experiment, you look at the two jars every day. You observe that the bean in the jar with water takes in water and gets bigger. Then you observe this bean growing into a plant. You also see that the bean in the jar with no water does not grow into a plant.

◀ The bean in the jar with water gets bigger and begins to grow.

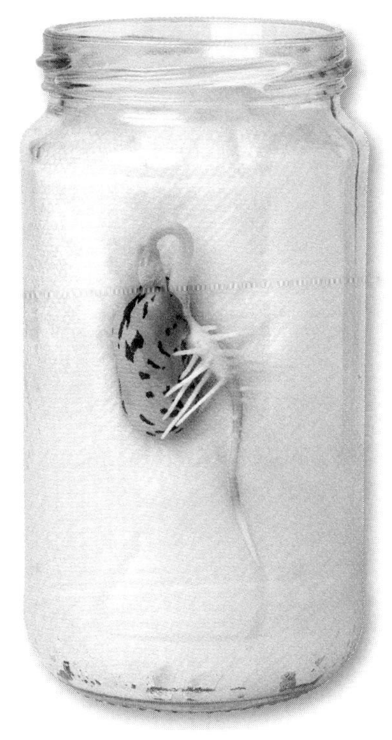

It is changing into a plant. ▶

Before You Go On

1. What do scientists do after they make a hypothesis?
2. What's a good way to test a hypothesis?
3. How do scientists observe?

Drawing Conclusions

After scientists observe, they decide if their hypothesis is correct. This is called drawing conclusions.

The bean experiment tested the hypothesis *seeds need water to grow.* You observed that the beans grew with water but not without water. The hypothesis is correct. Your conclusion is *seeds need water to grow.*

▲ This scientist is observing plants growing in pots.

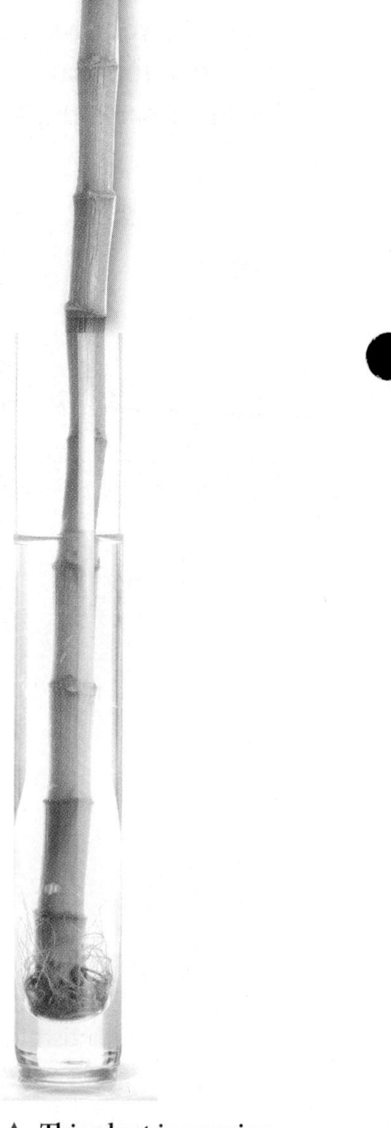

▲ This plant is growing in water. Plants need water to grow.

You can ask more questions about the bean plant. Will the plant continue to grow? Will it be a healthy plant? Does it need something else to grow well?

A scientist continues to observe and ask questions. A scientist thinks of another hypothesis and tests it. Then he or she draws other conclusions. Using the scientific method helps us learn about the world.

Scientific Method

Step 1: Ask questions.

↓

Step 2: Make a hypothesis.

↓

Step 3: Test the hypothesis.

↓

Step 4: Observe.

↓

Step 5: Draw conclusions.

▲ This scientist is studying a flower.

Before You Go On

1. What do scientists do after they observe?
2. What is the conclusion of the bean experiment?
3. What other questions can you ask about the bean plant?

For more practice, go to pages 33–34. 

Safety

Safety is very important in the science classroom. Always follow these basic safety rules. They will keep you and your classmates safe.

You will do an experiment to help you understand the scientific method. But first, learn these basic safety rules.

Make Sure You Understand

Read experiment instructions carefully. Make sure you understand before you begin. Ask your teacher when you don't understand.

Be Careful with Scissors

Always point scissors away from your body. When you carry scissors, point them down. Keep your fingers far away from scissor blades.

Stay Away from Broken Glass

If glass breaks, tell your teacher. Don't pick it up yourself.

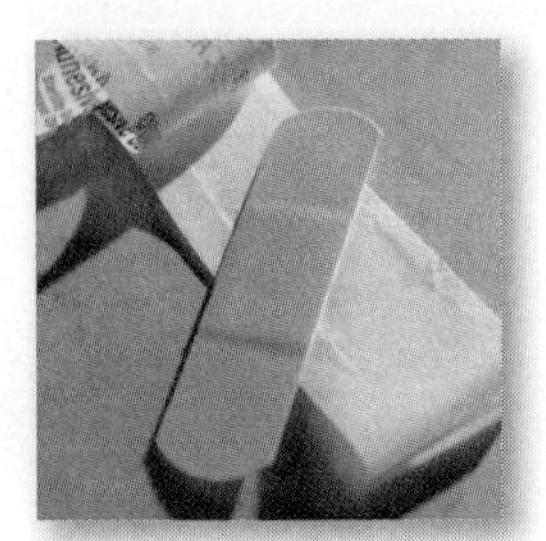

Tell Your Teacher If You Hurt Yourself

If you cut or hurt yourself, tell your teacher right away.

Clean Up Spills

Clean up any spills right away. Tell your teacher if anything spills on the floor.

Be Careful with Electricity

Be careful with things that use electricity. Don't use electrical items near water. Make sure the cords are out of the way.

Be Careful with Hot Things

Be careful with anything that is hot. Don't touch it until you know it is cool.

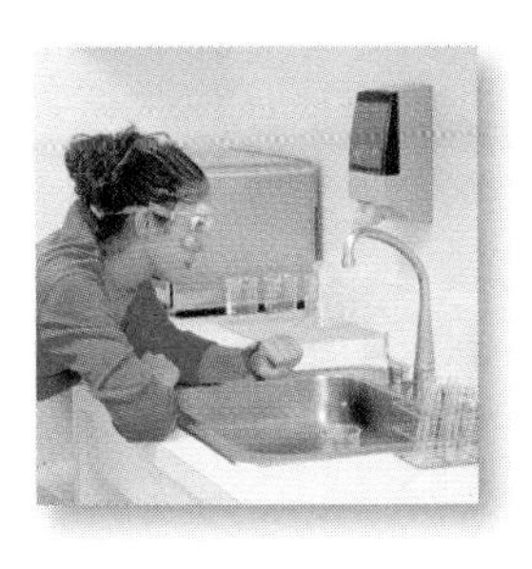

Keep Things Clean

Keep the experiment area clean and neat. Always put things away after you finish an experiment.

Before You Go On

1. Why do you follow safety rules?
2. What do you need to be careful with?
3. How do you carry scissors?

For more practice, go to pages 35–36. 

Practicing the Scientific Method

Practice using the scientific method as a class. Follow each step as you do the experiment on page 17.

1. Ask a question. Does air have weight?
2. Make a hypothesis. If you balance a stick with an equal-sized air-filled balloon attached to each end, will one side of the stick go up higher than the other? Using what you know, try to guess the answer. Your guess is your hypothesis.
3. Test the hypothesis. Do the experiment on the next page. It will test your hypothesis.

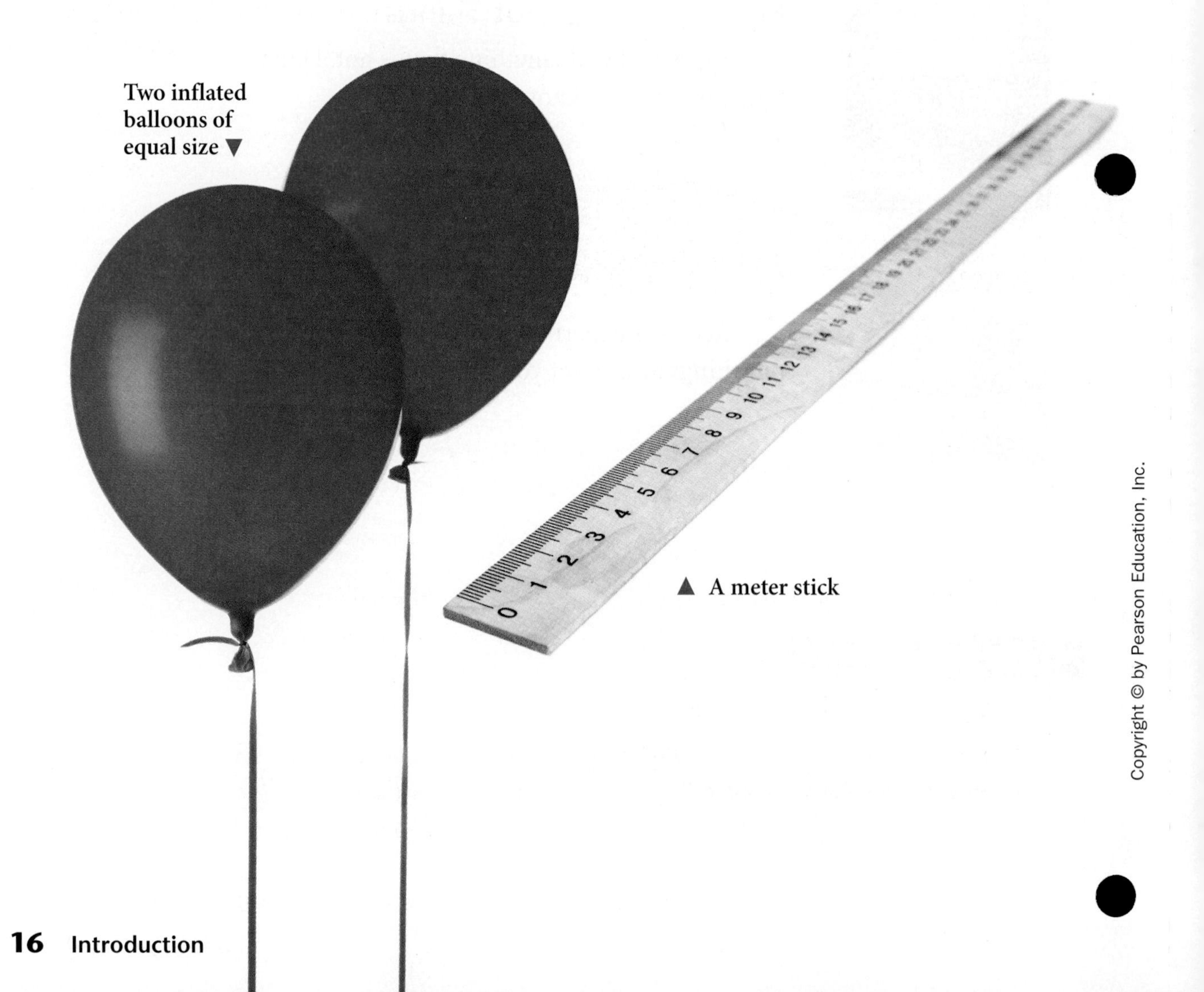

Two inflated balloons of equal size ▼

▲ A meter stick

Experiment

Does Air Have Weight?

Purpose

To find out if air has weight

Materials

two air-filled balloons of the same size and shape
a meter stick
tape or string to attach the balloons to the meter stick
a pin

What to Do

1. Fasten each air-filled balloon to one of the ends of the stick with a small piece of tape or string.
2. Find the center of the stick and place a finger there to balance the balloons. You may need to adjust the position of your finger to make the stick balance.
3. Once it is balanced, have a classmate pop one balloon with a pin.

Draw Conclusions

What happens when one balloon is popped? Why?

4. **Observe.** Which side of the balance went up when the balloon was popped? Why did this happen? In your Experiment Log, draw a picture of the meter stick before and after the balloon was popped.
5. **Draw conclusions.** What did you learn by observing the balance before and after one balloon was popped? Does an air-filled balloon weigh the same as one that is empty? Does this answer the question, "Does air have weight?" Your answer is your conclusion. Talk about what you learned with your classmates.

Experiment Log: Does Air Have Weight?

Follow the steps of the scientific method as you do your experiment.
Write notes about each step as the experiment progresses.

Step 1: Ask questions.

Step 2: Make a hypothesis.

Draw a picture here of the meter stick with the air-filled balloons attached before one balloon is popped.

Step 3: Test your hypothesis.

Step 4: Observe.

Draw a picture here of the meter stick with the balloons after one has been popped.

Step 5: Draw conclusions.

Science Tools

How hot is it? How fast is it? How big is it? How heavy is it? The tools on this page help scientists measure things. You will use some of these tools as you study science.

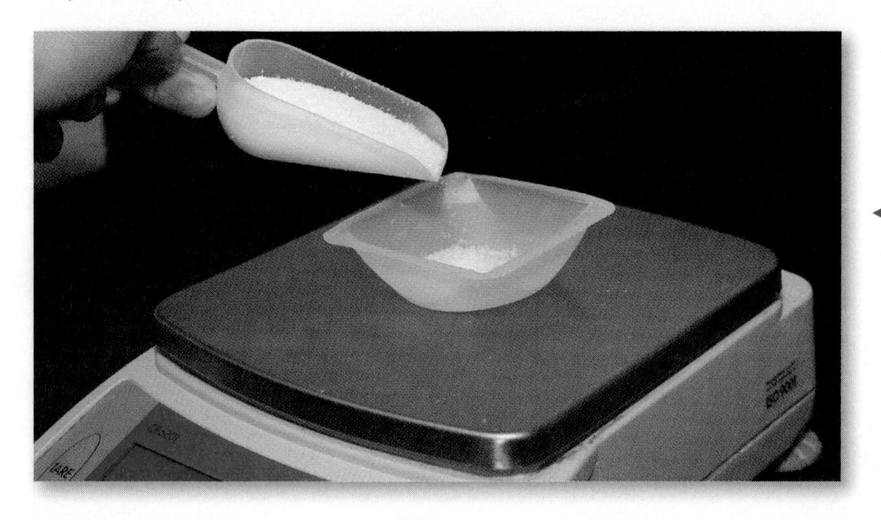

◀ A **balance** measures how heavy something is.

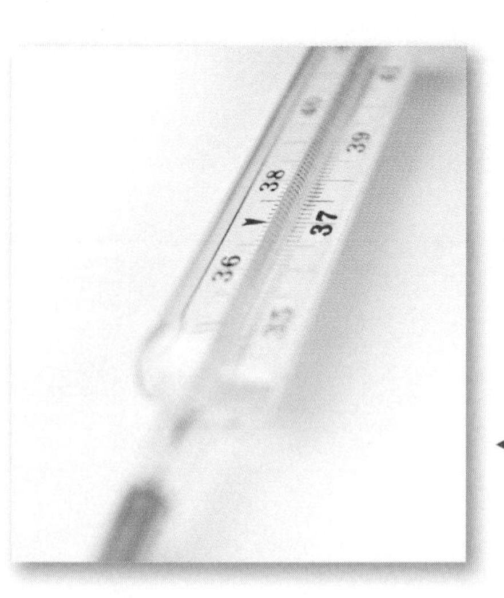

◀ A **thermometer** measures how hot or cold something is.

A **stopwatch** measures time. ▶

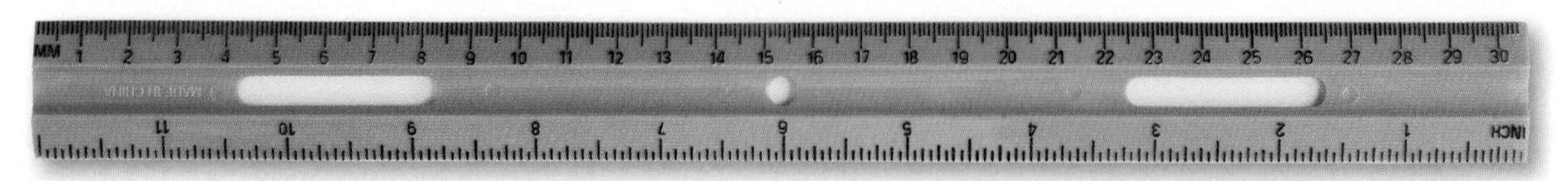

▲ A **ruler** measures distance. It helps us measure how long something is. This ruler shows both centimeters and inches. Scientists usually use centimeters.

How do you see something that is very small? How do you see a star far away in the sky? Scientists often use lenses made of glass. The tools on this page all have lenses.

▲ A **hand lens** lets us see small things close up.

▲ A **camera** lets us take pictures.

▲ A **telescope** helps us see distant stars.

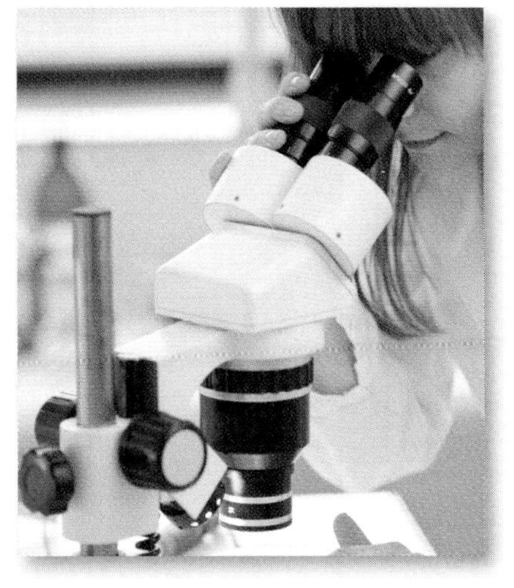

▲ A **microscope** lets us see very small things.

Before You Go On

1. What tool measures how hot or cold something is?
2. What four tools have lenses?
3. Which tool lets us see things far away in the sky?

For more practice, go to pages 37–38.

Visuals

After scientists learn information, they tell others about it. Scientists often use visuals to share information. These are some of the visuals you will study in this book.

Sound	Volume in Decibels
Shuttle taking off	190
Rock concert	115
Busy street	70
Classroom or office	45
Falling leaves	10

▲ Chart

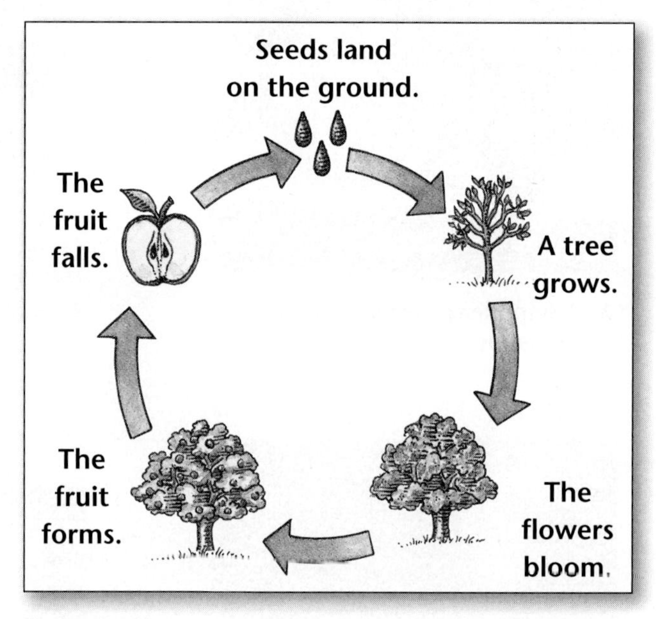

▲ Cycle diagram

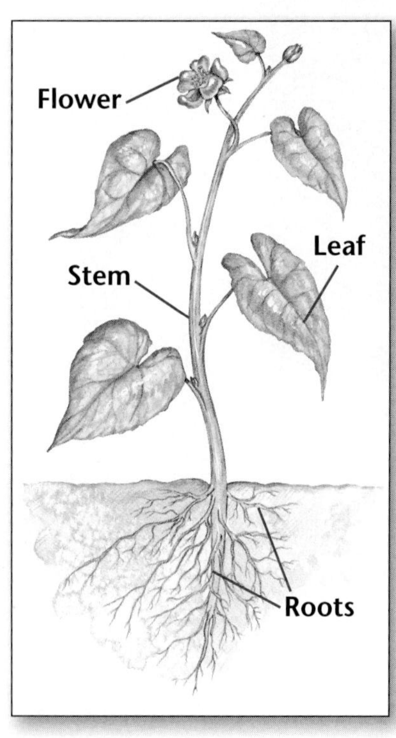

▲ Diagram

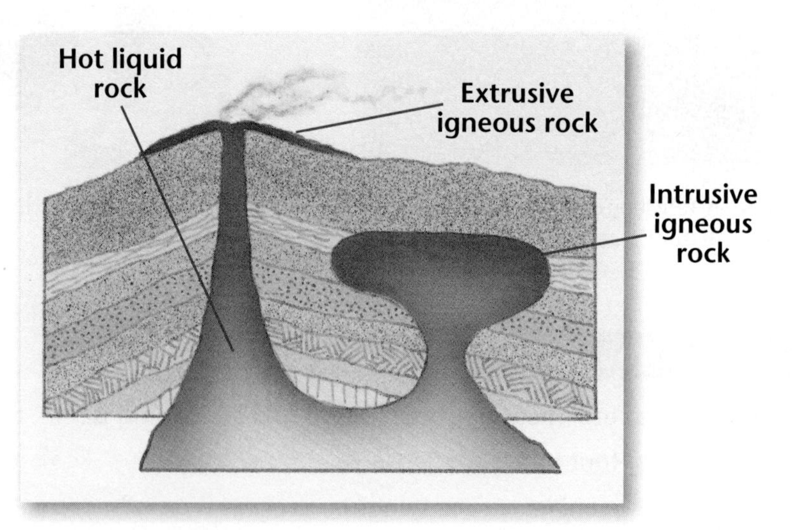

▲ Sectional diagram

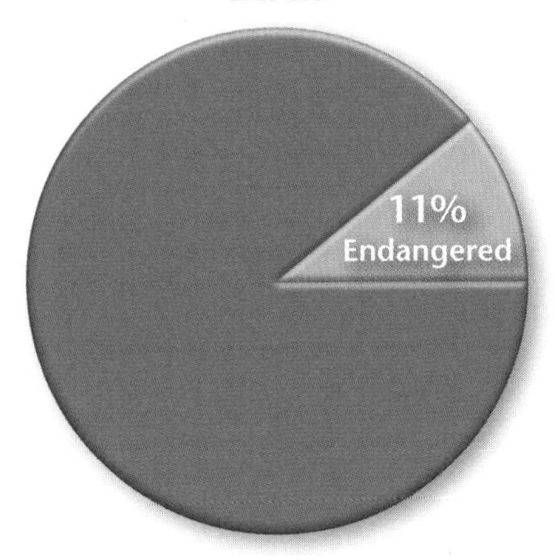

▲ Pie chart

▲ Illustration

▲ Photograph

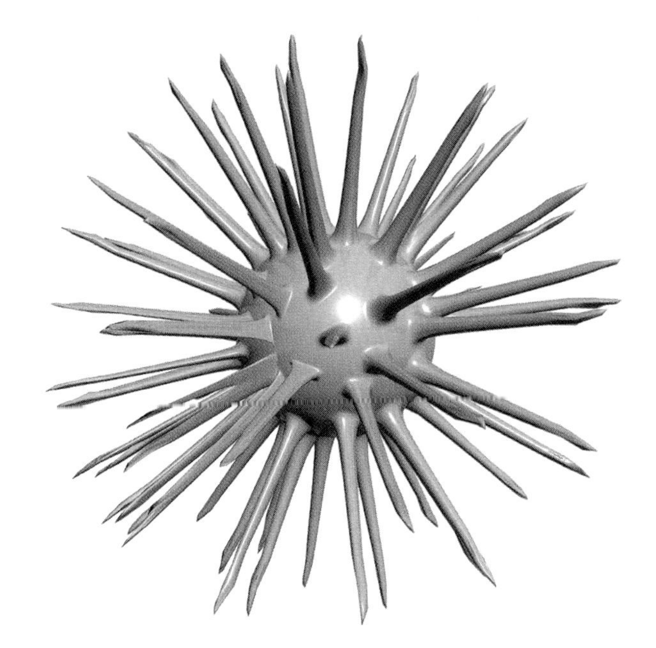

▲ Micrograph

Before You Go On

1. What do scientists use visuals for?
2. Which visuals give information using words and numbers?
3. Which visuals use pictures?

For more practice, go to pages 39–40.

Science Reading Strategies

In this book, you'll use science reading strategies. These strategies are special ways to look at and think about the text in your science book. These strategies will help you understand and remember what you read.

◀ **Good readers use reading strategies.**

Preview

Previewing is a reading strategy. When you preview, you look at pages before you read them. You look at the headings—the big or boldface words on the page. You also look at the pictures and the words near them.

Preview ▶

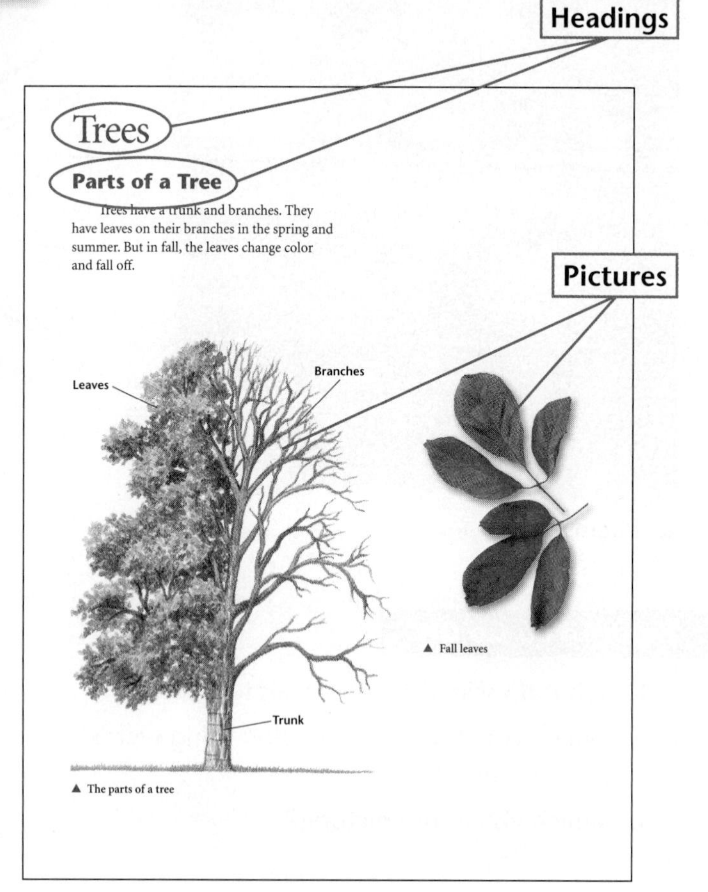

Predict

Another good reading strategy is predicting. To predict means to guess what a text is about. While you preview, predict what you will learn about.

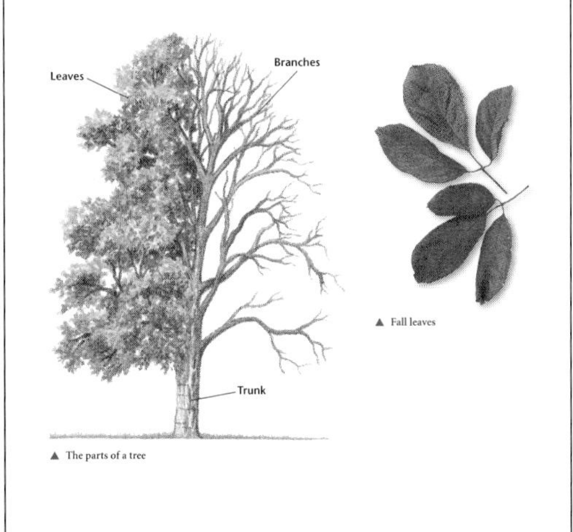
Trees

Parts of a Tree

Trees have a trunk and branches. They have leaves on their branches in the spring and summer. But in fall, the leaves change color and fall off.

▲ The parts of a tree

▲ Fall leaves

▲ Predict

▲ Most people preview and predict when they first look at a magazine.

Before You Go On

1. How are science reading strategies helpful?
2. How do you preview?
3. What does *predict* mean?

Getting Started

Review and Practice

Check Your Understanding

Complete each sentence with the correct word or phrase.

1. ______________ science is the study of the Earth.
   a. Physical b. Life c. Earth

2. ______________ science is the study of living things.
   a. Earth b. Life c. Physical

3. ______________ science is the study of matter and energy.
   a. Life b. Physical c. Earth

4. Asking questions and making a hypothesis are steps in the ______________.
   a. visuals b. science tools c. scientific method

5. Doing an experiment is a good way to test a ______________.
   a. hypothesis b. conclusion c. safety rule

6. Scientists ______________ by watching very carefully.
   a. guess b. observe c. ask

7. After an experiment, scientists draw ______________.
   a. conclusions b. questions c. safety rules

8. A thermometer, a telescope, and a balance are science ______________.
   a. visuals b. tools c. strategies

9. A diagram, a pie chart, and a micrograph are ______________.
   a. tools b. strategies c. visuals

10. Previewing and predicting are reading ______________.
    a. strategies b. tools c. visuals

Apply Science Skills

Science Reading Strategy: **Preview and Predict**

As a class, preview these pages from a life science book. As you preview, predict. Tell your teacher what you will learn about.

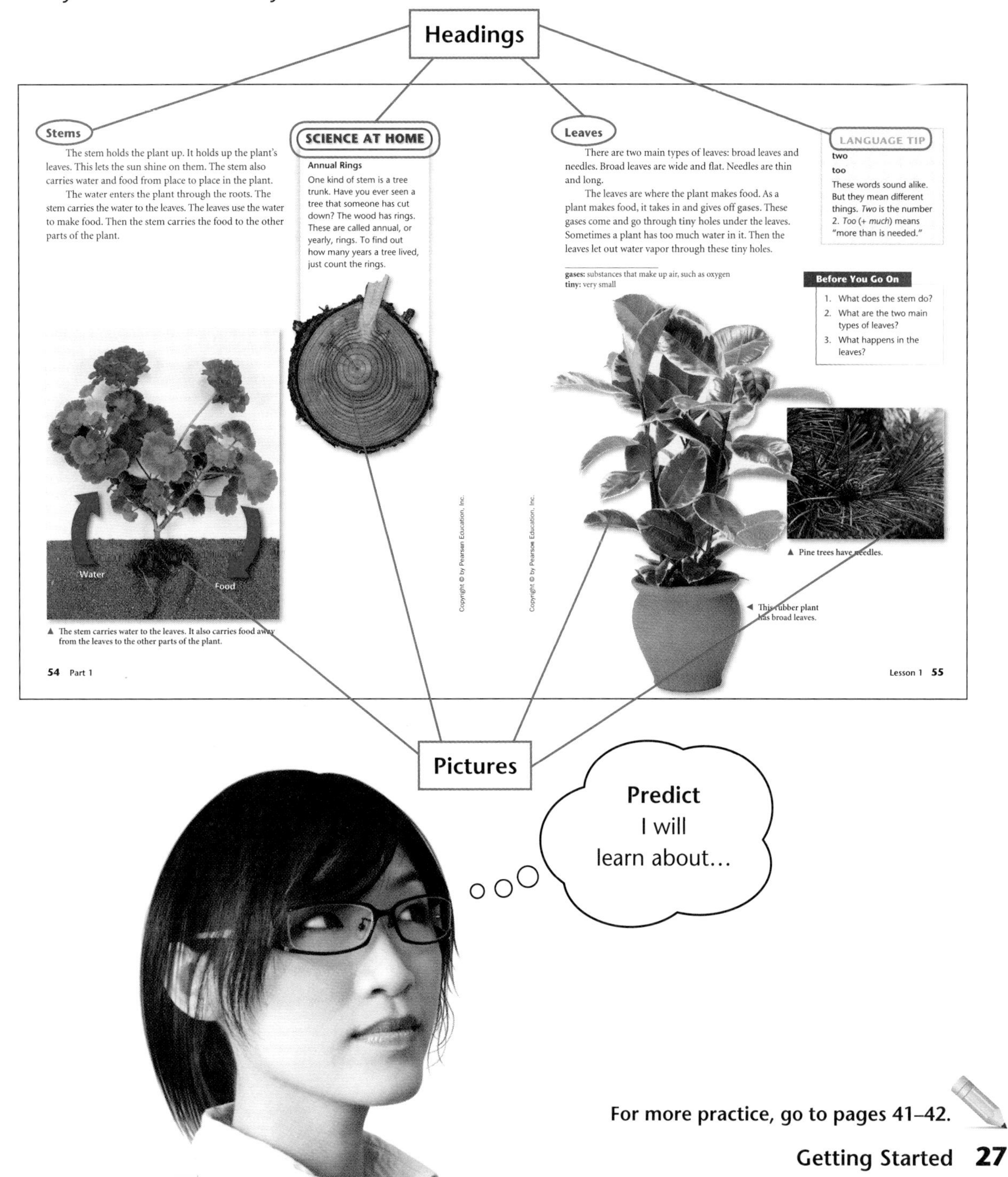

Stems

The stem holds the plant up. It holds up the plant's leaves. This lets the sun shine on them. The stem also carries water and food from place to place in the plant.

The water enters the plant through the roots. The stem carries the water to the leaves. The leaves use the water to make food. Then the stem carries the food to the other parts of the plant.

SCIENCE AT HOME

Annual Rings

One kind of stem is a tree trunk. Have you ever seen a tree that someone has cut down? The wood has rings. These are called annual, or yearly, rings. To find out how many years a tree lived, just count the rings.

Water

Food

▲ The stem carries water to the leaves. It also carries food away from the leaves to the other parts of the plant.

54 Part 1

Leaves

There are two main types of leaves: broad leaves and needles. Broad leaves are wide and flat. Needles are thin and long.

The leaves are where the plant makes food. As a plant makes food, it takes in and gives off gases. These gases come and go through tiny holes under the leaves. Sometimes a plant has too much water in it. Then the leaves let out water vapor through these tiny holes.

gases: substances that make up air, such as oxygen
tiny: very small

LANGUAGE TIP

two

too

These words sound alike. But they mean different things. *Two* is the number 2. *Too* (+ *much*) means "more than is needed."

Before You Go On

1. What does the stem do?
2. What are the two main types of leaves?
3. What happens in the leaves?

▲ Pine trees have needles.

◄ This rubber plant has broad leaves.

Lesson 1 55

For more practice, go to pages 41–42.

Practice Pages

Name ______________________________ Date ______________

What Is Science?

A. Match the parts of the sentence. Write the letter.

c	**1.** Science	**a.**	is the Earth and all things on it.
______	**2.** Scientists	**b.**	are things that are alive.
______	**3.** Living things	**c.**	is the study of the natural world.
______	**4.** Nonliving things	**d.**	are people who study our world.
______	**5.** The world	**e.**	are things that are not alive.

B. Complete each sentence. Use words from the box.

1. ____________________ is the Earth and all things on it.

2. ____________________ are people who study our world.

3. ____________________ is the study of the natural world.

4. ____________________ are things that are not alive.

5. ____________________ are things that are alive.

Living things
Nonliving things
Science
Scientists
The world

C. Write *living thing* or *nonliving thing* under each picture.

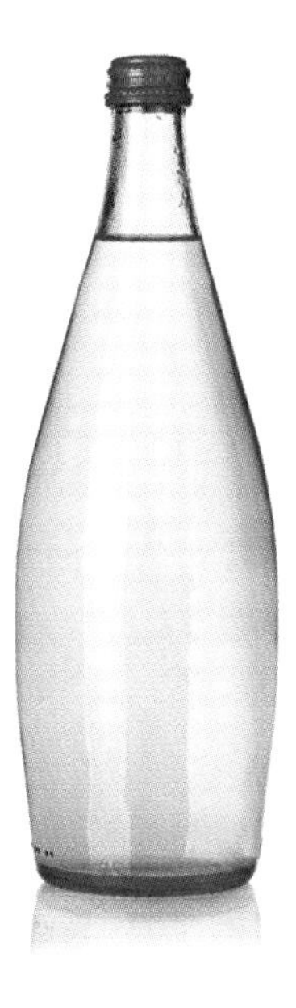

PRACTICE

The Sciences

A. Match the parts of the sentence. Write the letter.

d	Life science	**a.** is the study of mostly nonliving matter.
______	**1.** Earth science	**b.** is what living and nonliving things are made of.
______	**2.** The environment	**c.** is power such as sound, light, and electricity.
______	**3.** Matter	**d.** is the study of living things.
______	**4.** Physical science	**e.** is the land, water, air, and living things on the Earth.
______	**5.** Energy	**f.** is the study of the Earth.

B. Write five sentences with words and phrases from the exercise above.

Example: *Life science is the study of living things.*

1. ______________________________

2. ______________________________

3. ______________________________

4. ______________________________

5. ______________________________

C. Circle the word or phrase that doesn't belong.

1. electricity	energy	(plants)	physical science
2. life science	rocks	animals	plants
3. animals	air	rocks	land
4. Earth science	water	planets	air
5. energy	sound	light	rocks
6. humans	electricity	frogs	trees

Name ______________________________ Date ______________

D. Under each picture, write the correct kind of science. Choose words from the box.

Earth science	physical science	life science

______________ ______________ ______________

______________ ______________ ______________

PRACTICE

E. Complete the chart below. Write an example of each kind of science.

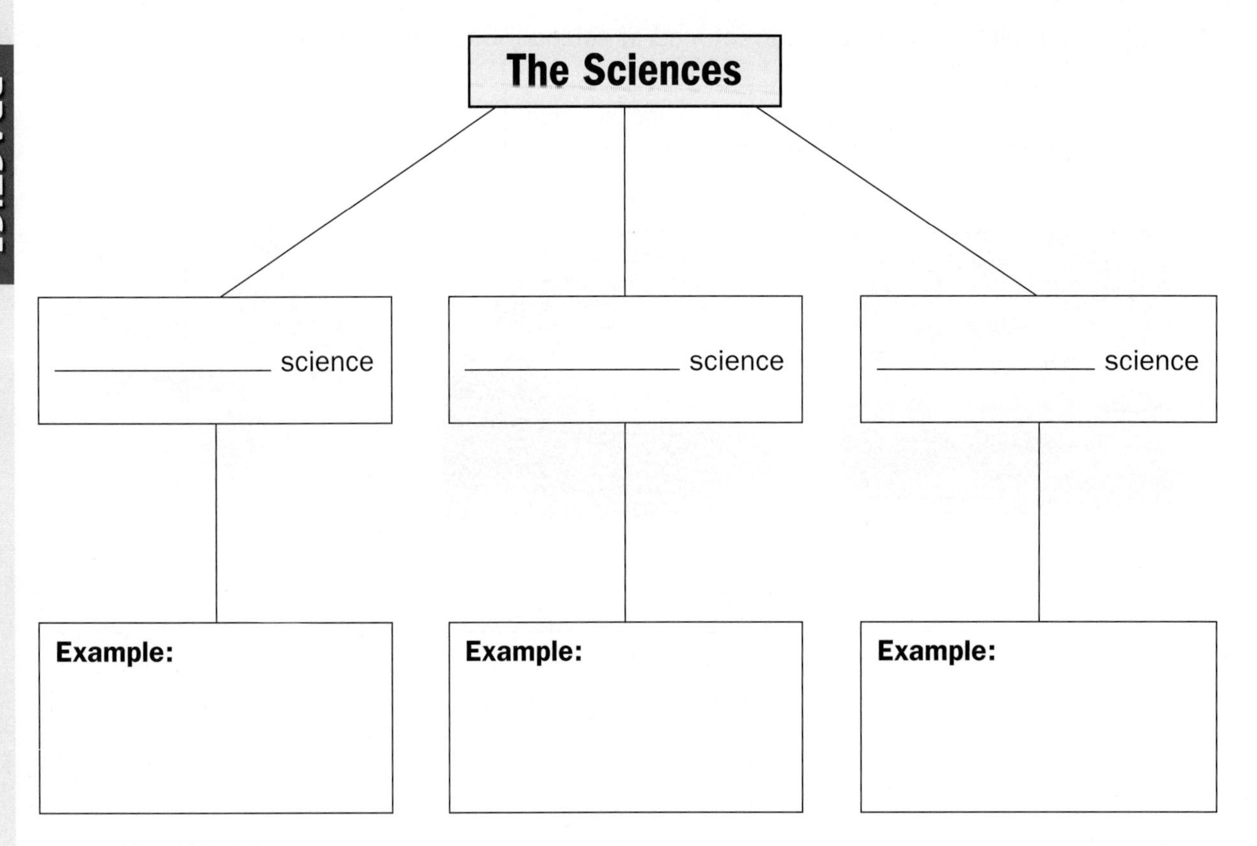

F. Choose the best answer. Circle the letter.

1. _______ is the study of living things on the Earth.

a. Earth science **b.** Physical science **c.** Life science

2. The study of energy is part of _______.

a. Earth science **b.** physical science **c.** life science

3. Our _______ is the land, water, air, and living things around us.

a. living thing **b.** environment **c.** sun

4. _______ is the study of the Earth.

a. Earth science **b.** Physical science **c.** Life science

5. _______ is the study of mostly nonliving matter.

a. Earth science **b.** Physical science **c.** Life science

Name ______________________ Date __________

The Scientific Method

PRACTICE

A. Write the steps of the scientific method in the correct order. Use the sentences from the box.

- Draw conclusions.
- Ask questions.
- Test the hypothesis.
- Make a hypothesis.
- Observe.

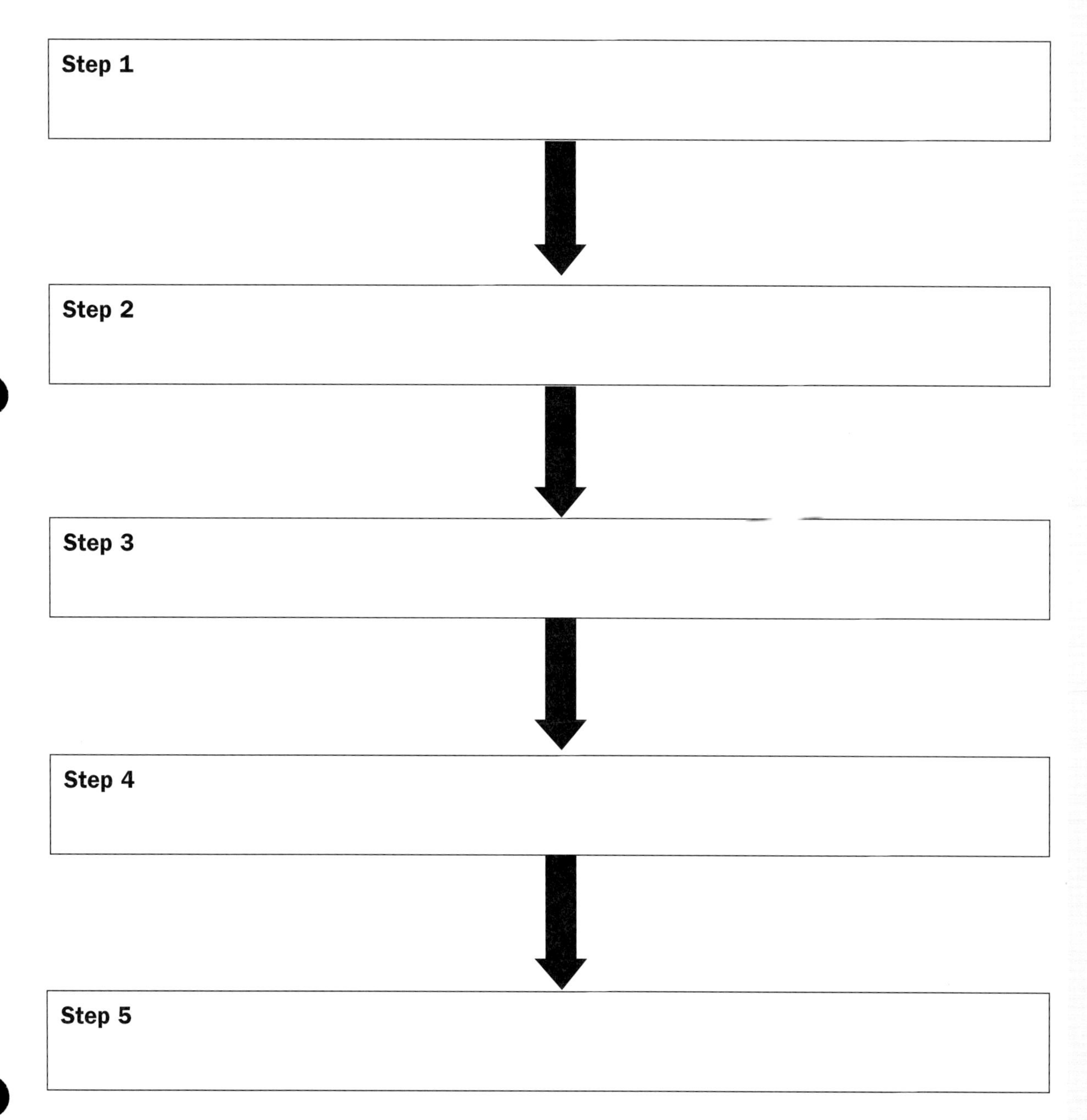

PRACTICE

B. Match the parts of the sentence. Write the letter.

_______ **1.** First, scientists ask questions

_______ **2.** After asking a question,

_______ **3.** After scientists make a hypothesis,

_______ **4.** Scientists observe

_______ **5.** Finally, when scientists draw conclusions,

a. scientists make a hypothesis.

b. very carefully to see what happens.

c. they decide if their hypothesis is correct.

d. about things they don't know.

e. they test their hypothesis.

C. Write five sentences with phrases from the exercise above.

1. ______________________________

2. ______________________________

3. ______________________________

4. ______________________________

5. ______________________________

D. Write T for *true* or F for *false*.

_______ **1.** Scientists don't ask questions.

_______ **2.** A hypothesis is a guess.

_______ **3.** Scientists do experiments to test a hypothesis.

_______ **4.** First scientists draw conclusions, then they make a hypothesis.

_______ **5.** Observing is not part of the scientific method.

Name ______________________ Date ____________

Safety

A. Match each safety rule with a picture. Write the letter.

Safety Rules

a. Clean up spills.

b. Be careful with scissors.

c. Make sure you understand.

d. Be careful with electricity.

e. Be careful with hot things.

f. Keep things clean.

g. Stay away from broken glass.

h. Tell your teacher if you hurt yourself.

______

______

______

______

______

______

______

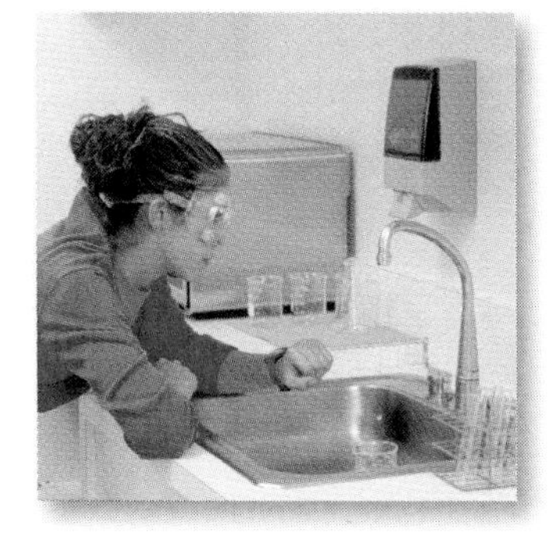

______

B. Match the parts of the sentence. Write the letter.

_______	**1.** Keep	**a.**	if anything spills on the floor.
_______	**2.** Stay away	**b.**	with scissors.
_______	**3.** Tell your teacher	**c.**	electrical cords are out of the way.
_______	**4.** Make sure	**d.**	things clean.
_______	**5.** Be careful	**e.**	from broken glass.

C. Write T for *true* or F for *false*.

_______ **1.** Always point scissors away from your body.

_______ **2.** Pick up broken glass yourself.

_______ **3.** Don't put things away after an experiment.

_______ **4.** Don't use electrical items near water.

_______ **5.** Don't clean up spills.

D. Complete the paragraph. Use words from the box.

clean	teacher	understand	hot	electricity

Safety is very important in the science classroom. You should learn these basic safety rules. Make sure you **(1)** ____________________ the experiment before you begin. Be careful with **(2)** ____________________ things. Be careful with **(3)** ____________________, too. Make sure the cords are out of the way. Keep the experiment area **(4)** ____________________. Tell your **(5)** ____________________ if you hurt yourself.

Name ______________________ Date ____________

Science Tools

PRACTICE

VOCABULARY

A. Write the correct word under each picture. Use words from the box.

camera	hand lens	stopwatch
thermometer	microscope	balance

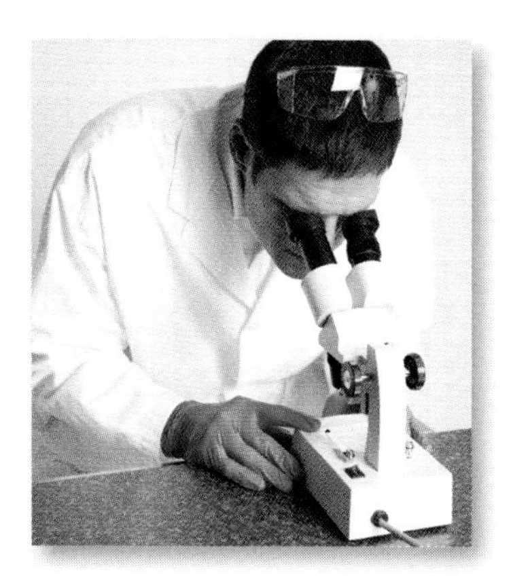

1. ____________

2. ____________

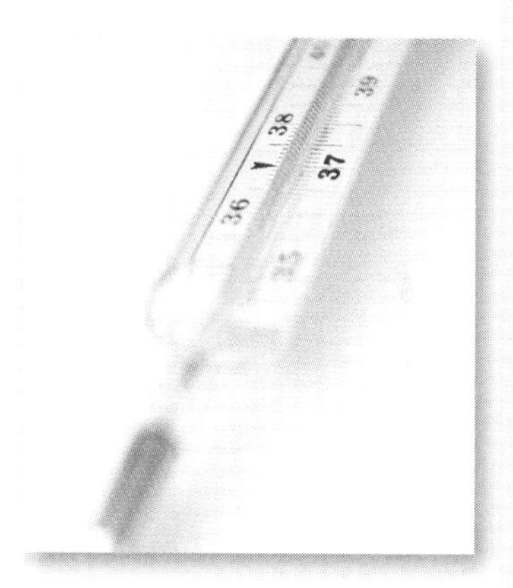

3. ____________

4. ____________

5. ____________

6. ____________

PRACTICE

B. Match the parts of the sentence. Write the letter.

_______	**1.** A ruler	**a.** lets you see small things close up.
_______	**2.** A hand lens	**b.** lets you see very small things.
_______	**3.** A balance	**c.** lets you measure how hot or cold something is.
_______	**4.** A telescope	**d.** lets you take pictures.
_______	**5.** A stopwatch	**e.** lets you measure how long something is.
_______	**6.** A thermometer	**f.** lets you see things far away.
_______	**7.** A camera	**g.** lets you measure how heavy something is.
_______	**8.** A microscope	**h.** lets you measure time.

C. Write eight sentences with words and phrases from the exercise above.

1. ______________________________

2. ______________________________

3. ______________________________

4. ______________________________

5. ______________________________

6. ______________________________

7. ______________________________

8. ______________________________

D. Complete each sentence. Use words from the box.

thermometer	microscope	telescope	hand lens	ruler

1. A ______________ has centimeters and inches.

2. You hold a ______________ in your hand to observe small things.

3. A ______________ measures how hot or cold something is.

4. You look through a ______________ to see things that are far away.

5. You can observe very, very small things with a ______________.

Name ______________________ Date ____________

Visuals

Write the name of the visual under each picture. Use words or phrases from the box.

photograph	cycle diagram	diagram	pie chart
illustration	micrograph	chart	sectional diagram

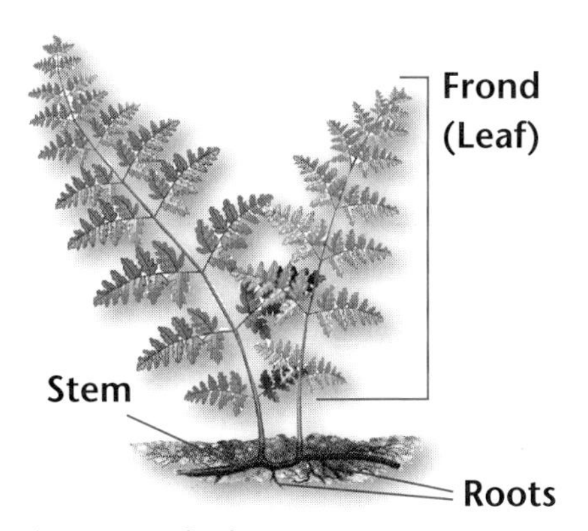

▲ Parts of a fern

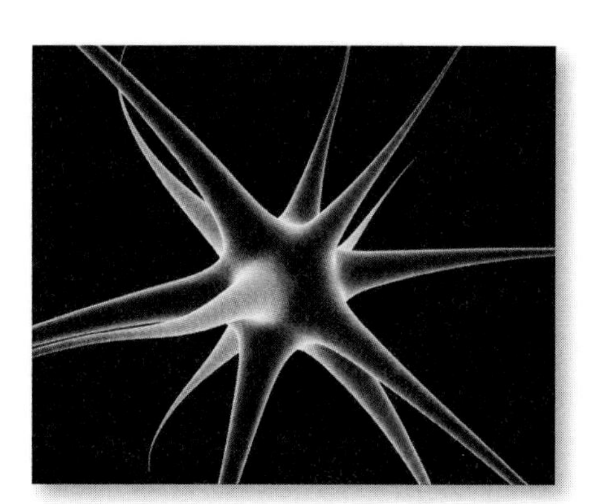

1. ______________________

2. ______________________

3. ______________________

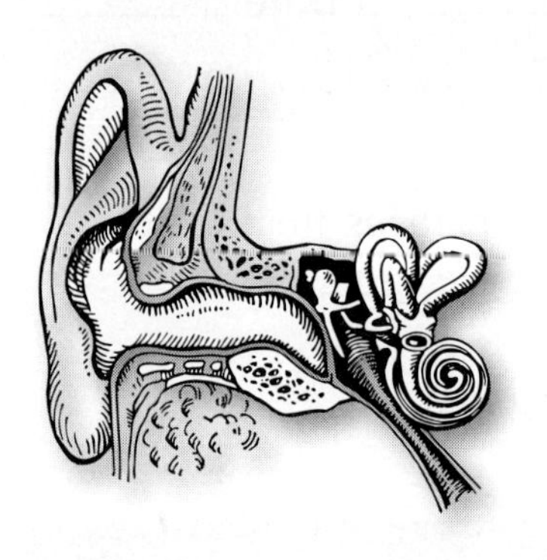

4. ______________________

5. ______________________

Matter	Speed of Sound	
	Meters per second	Feet per second
Dry, cold air	343	1,125
Water	1,550	5,085
Hard wood	3,960	12,992
Glass	4,540	14,895
Steel	5,050	16,568

6. ______________________

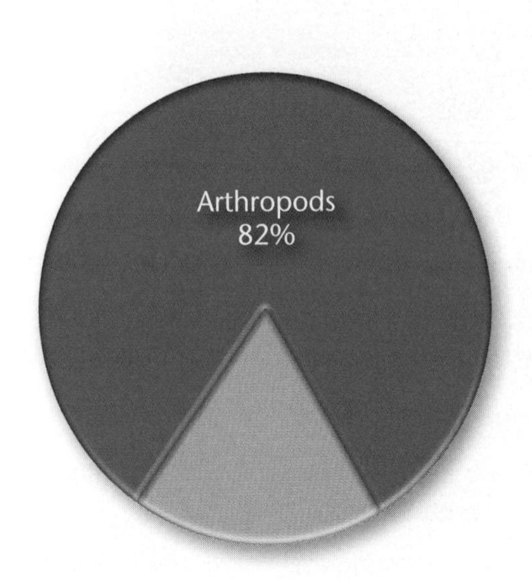

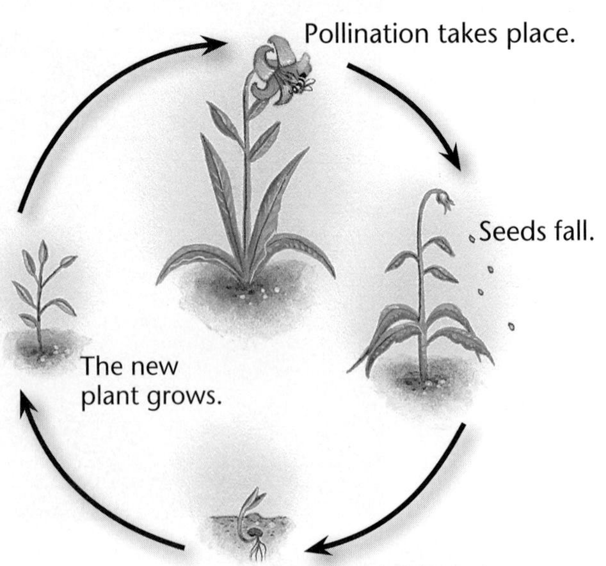

7. ______________________

8. ______________________

Review

VOCABULARY

Complete the puzzle. Use words from the box.

balance	microscope
energy	nonliving things
environment	science
hypothesis	stopwatch
life science	world
matter	

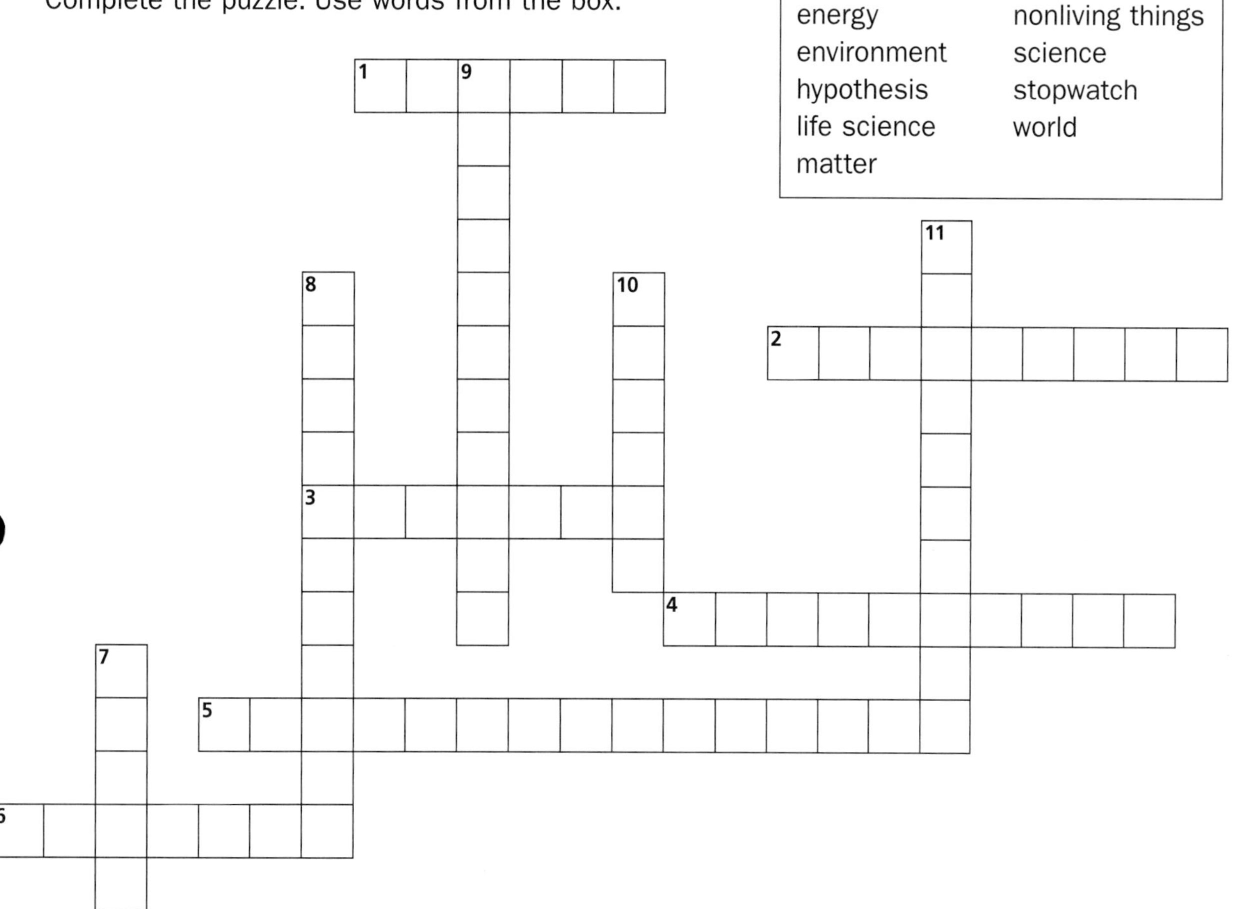

ACROSS

1. power such as electricity
2. measures time
3. the study of the natural world
4. lets you see very small things
5. things that are not alive
6. measures how heavy something is

DOWN

7. Earth and all things on it
8. the study of living things
9. land, water, air, and living things on the Earth
10. what living and nonliving things are made of
11. a scientist's guess

CHECK YOUR UNDERSTANDING

Write T for *true* or F for *false*.

_______ **1.** Scientists use visuals to share information.

_______ **2.** Physical science is the study of living matter.

_______ **3.** Never tell your teacher if you hurt yourself.

_______ **4.** Life science is the study of living things on the Earth.

_______ **5.** The scientific method has five steps.

_______ **6.** A balance measures how hot something is.

_______ **7.** Safety is not important in the science classroom.

_______ **8.** You draw conclusions at the end of an experiment.

_______ **9.** Earth science includes the study of our environment.

_______ **10.** A hypothesis is always correct.

APPLY SCIENCE SKILLS

Science Reading Strategy: **Preview and Predict**

Look at the pictures on page 2. Then answer the questions.

1. What do the pictures show?

____

2. Which picture shows living things?

____

3. Which picture shows nonliving things?

____

4. Predict what you will learn about in your science book.

____

____

____

Getting Started

Science Journal

Write about five interesting things you have learned in Getting Started.

1. ______________________________

____

____

2. ______________________________

____

____

3. ______________________________

____

____

4. ______________________________

____

____

5. ______________________________

____

____

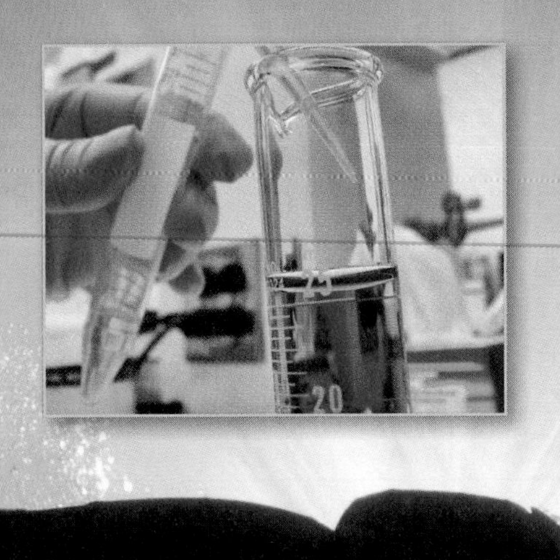

Matter

Part 1

Part Concepts

Lesson 1

- Matter is anything that takes up space and has mass.
- Each type of matter has its own properties.
- You use your senses to learn about some properties.
- You measure matter to learn about other properties.
- Matter can be a pure **substance** or a mixture.
- All matter is made up of tiny **particles**, or pieces, called atoms.

Lesson 2

- There are three states of matter: solid, liquid, and gas.
- Matter can change from one state to another.
- Melting point is the temperature at which a solid changes to a liquid.
- Boiling point is the temperature at which a liquid changes to a gas.
- In a physical change, matter changes how it looks.
- In a chemical change, the groups of atoms that make up matter change.

Get Ready

Everything in your classroom is a kind of matter. This chart lists different kinds of matter. Find objects in the classroom made of each kind of matter. Write the names of the objects in the chart.

Kinds of Matter	Objects
wood	
metal	
plastic	
glass	

substance: physical matter or material
particle: a very small piece of matter

Lesson 1 —Before You Read

Vocabulary

▲ Everything you see in this picture is **matter**. Matter is anything that takes up space and has mass. Mass is like weight.

You can measure the **mass** of something with a balance like this one. Mass is how much of something there is. It is one of the object's **properties**. ▼

▲ You can use a measuring cup to measure **volume**. Volume is how much space something takes up and is another property of matter.

The volume of the gases inside these balloons is the same. The gas inside the balloon on the left has more mass, so it has greater **density**. Density is another of matter's properties. ▼

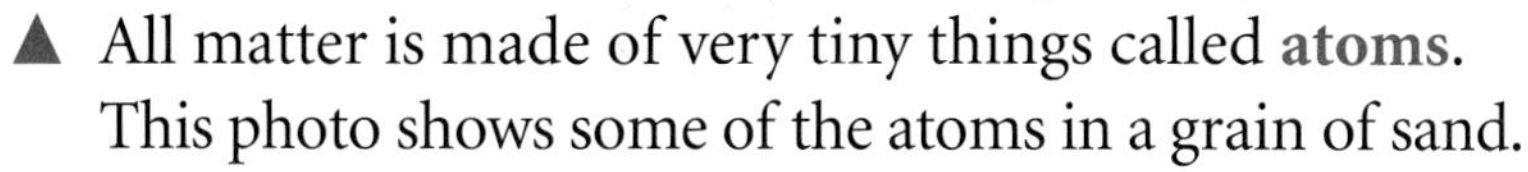

▲ All matter is made of very tiny things called **atoms**. This photo shows some of the atoms in a grain of sand.

Key Words

atoms

density

mass

matter

properties

volume

Practice

Choose the word that completes each sentence.

1. Matter is anything that takes up space and has ____________.
   a. volume **b.** mass **c.** measure
2. All matter is made of very tiny things called ____________.
   a. mass **b.** density **c.** atoms
3. Two things have the same volume. The thing with more mass has greater ________.
   a. density **b.** measure **c.** volume
4. A ____________ of matter is its mass.
   a. volume **b.** property **c.** density
5. ____________ is how much space something takes up.
   a. Mass **b.** Density **c.** Volume
6. Everything you see around you is ____________.
   a. matter **b.** mass **c.** measure

For more practice, go to page 61.

Science Skills

Science Reading Strategy: **Facts and Examples**

When you read, look for **facts and examples.**

- Facts are main science points. Examples help you understand facts.
- Look for the words *for example* or *suppose.*
- Look for examples in the pictures and captions, too.

Read this text. Look at diagram 1. Then complete diagram 2.

Helium is a gas. There is a lot of helium in the universe. But there is not much helium on Earth. Helium is lighter than air and will float above it. For example, a balloon filled with helium will rise. The helium in the balloon has lower density than air. In fact, helium has such low density that it quickly leaves Earth. Suppose you let go of a helium balloon. About five miles up in the sky, the balloon will break. Then the helium will go into space.

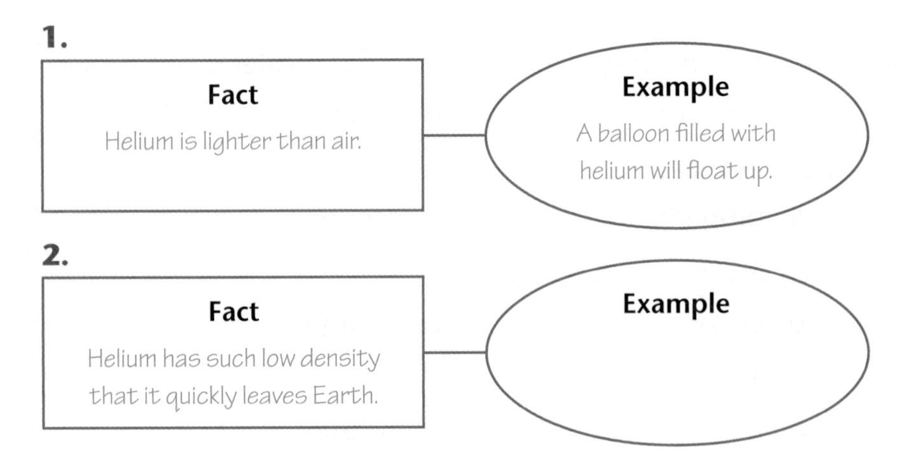

Look for an **As You Read** question about facts and examples in the lesson. It will help you understand the lesson.

For more practice, go to pages 62–63. 

Using Visuals: **Micrographs**

These pictures are micrographs. Micrographs are photographs taken with a microscope.

Look at the micrographs. Then answer the questions.

▲ Snowflake

▲ Sand

▲ Sugar

▲ Salt

1. Which substance is a mixture of different kinds of matter?
2. Which substances are made of only one kind of matter?
3. What is the same in all the substances?

For more practice, go to page 64.

Lesson 1

What Is Matter?

The word *space* can mean the area beyond the Earth's atmosphere. Here, *space* means the area around you.

Everything you see is matter. Matter can be living or nonliving. It can be a book, a cloud, or a person. All these things take up space. Matter is anything that takes up space and has mass. Mass is **similar** to weight. We talk about mass in terms of grams and kilograms. For example, 1 liter of water has a mass of 1 kilogram.

Properties

Each type of matter has its own properties. Color, size, and shape are properties you can see or feel.

similar: having a likeness or resemblance; alike

▲ Everything in this picture takes up space and has mass. Everything in this picture is matter.

▲ This is 1 liter of water. It has a mass of 1 kilogram.

For example, think about some of wood's properties. Most wood is hard. You can build things with it. Most types of wood float on water. Wood burns easily, so we use it as fuel.

Honey has a different set of properties. It is sweet and sticky. You can pour it. It does not have its own shape. It takes the shape of its container. Light shines through it. It **dissolves** in water.

---

dissolves: breaks into very tiny pieces in liquid

▲ **Honey is a liquid. It flows very slowly.**

▲ **This piece of wood is a good place for a bird to rest.**

Before You Go On

1. What is matter?
2. What are some properties of wood?
3. Can you think of other kinds of matter that dissolve in water? Name them.

You can use your senses to learn about some properties of matter. You can see the color and shape of a lemon. You can feel the bumps on its skin. You can taste how sour it is.

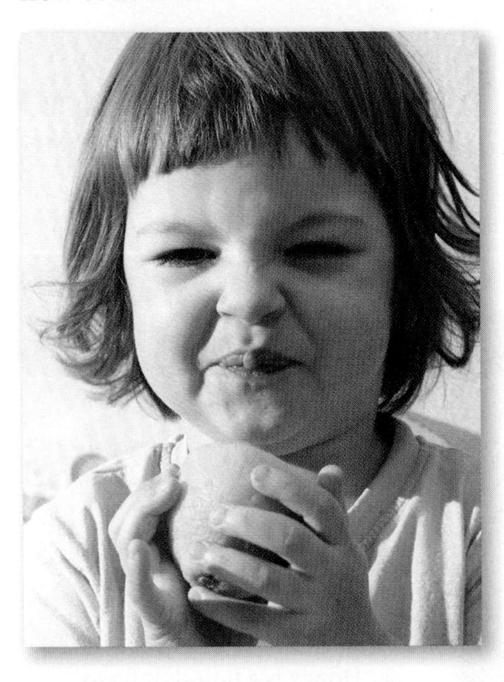

◀ Lemons taste sour.

SCIENCE AT HOME

Measuring Matter

You and your family measure matter every day. Go to the kitchen and bathroom in your home. Look for ways you measure matter.

▲ We often measure matter when we cook.

Mass

Mass is another property of matter. Mass is how much of something there is. You measure the mass of something by using a balance.

This balance is measuring the mass of a block of metal. ▼

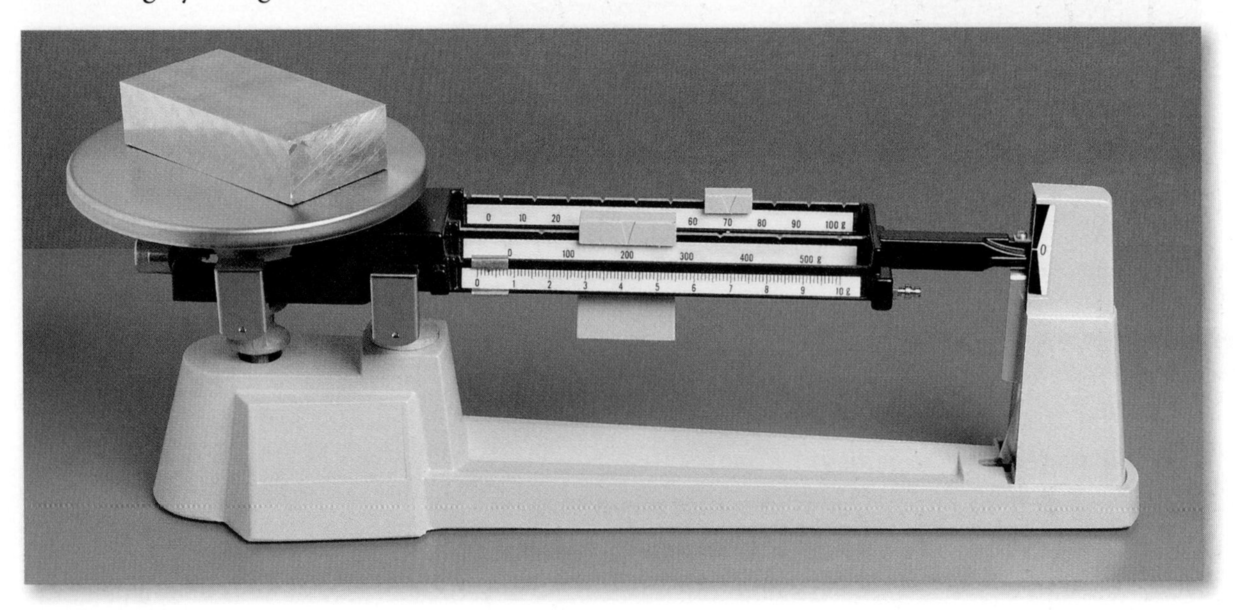

Volume

Another property that people measure is volume. Volume is the amount of space something takes up. You use containers of different sizes to measure volume. We talk about volume in terms of liters.

Think about buying water at the store. The water is measured for you. You can buy a 1-liter container or a 2-liter container. A 2-liter container has two times the volume of a 1-liter container.

LANGUAGE TIP

contain → verb
container → noun

A cup contains *water.*
A cup is a container.

Facts and Examples

Fact: You use containers of different sizes to measure volume. What examples can you find on this page?

▲ Water and other drinks come in 2-liter containers like this one.

A measuring cup measures volume. ▶

Before You Go On

1. What do you use to measure mass?
2. What is volume?
3. Which is easier to measure, the volume of water or the volume of a rock? Why?

Exploring

Measurement in History

The earliest unit of measurement was food. The ancient Egyptians used grains of wheat as weights. The smallest unit was one grain.

By 1400 B.C.E., people also used metal pieces to measure weight. In Egypt, people made metal weights shaped like animals or rings. The rings could also be used as money.

The Romans first used the pound as a unit of weight around 200 B.C.E. The abbreviation for pound is *lb*. This is really an abbreviation for *libra*. *Libra* means "balance" in Latin, the language of ancient Rome.

The metric system was invented in France in the 1790s. In this system, 1 liter of water equals 1 kilogram. The metric system was easy for people to use. Now it is used all over the world.

Density

A third property that we measure is density. Density is how much matter there is in a certain amount of space. It is how much mass something has per unit of volume.

Suppose you have two cups of the same size. You fill one cup with pennies and the other with puffed rice cereal. The cups have the same volume. Which cup has more mass?

A cup of pennies has more mass than a cup of puffed rice. That's another way of saying that pennies have greater density than puffed rice.

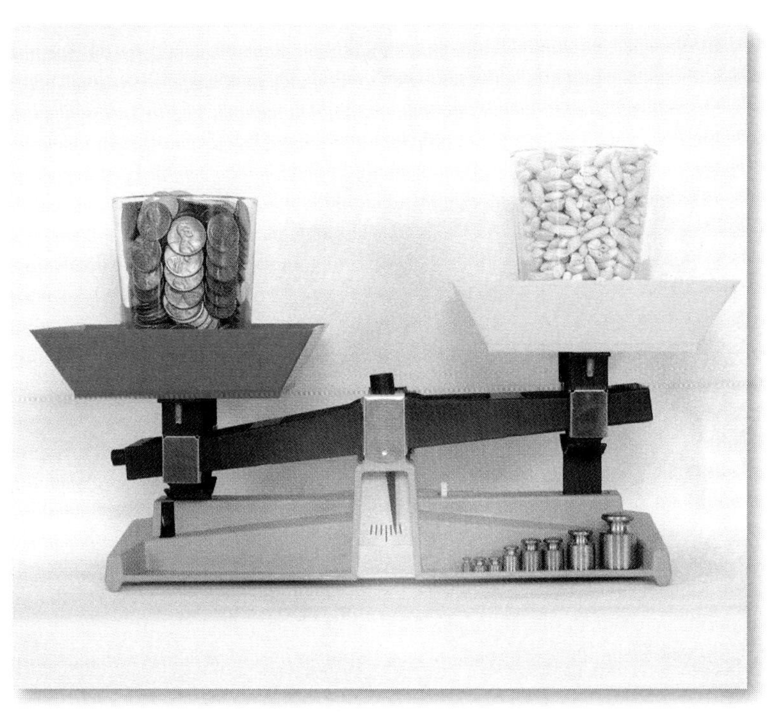

▲ **The pennies and puffed rice have the same volume. But the pennies have more mass. Pennies have greater density than puffed rice.**

Before You Go On

1. What is density?
2. Which has greater density, pennies or puffed rice?
3. Which do you think has greater density, a rock or a cotton ball?

LANGUAGE TIP

per unit → for each unit

percent → for each hundred

MATH CONNECTION

Measuring Density

You can measure the density of all kinds of matter. First you measure the volume and the mass of the substance. Then you divide the mass by the volume. For example, 1 liter of water weighs 1 kilogram. 1 divided by 1 equals 1. Water has a density of 1.

$$\text{Density} = \frac{\text{Mass}}{\text{Volume}}$$

What Makes Up Matter?

Matter can be a single, or pure, substance or a mixture of substances. A pure substance is made of only one kind of matter. Salt and sugar are examples of pure substances.

A mixture is made of different kinds of matter mixed together. Sand is an example of a mixture. Look at the micrograph of sand on page 49. You can see that it is made of different kinds of rock mixed together. Another example of a mixture is a salad.

▼ **Sugar is a pure substance.**

▲ **This salad is a mixture of different kinds of matter.**

All matter is made up of tiny pieces called atoms. Atoms are so small that 5 million of them fit on the head of a pin. You need a special microscope to see atoms.

The dots that make up this painting are like the atoms that make up all matter. ▼

▲ *A Sunday on La Grande Jatte,* 1884, by Georges Seurat

◀ This micrograph shows atoms in the metal nickel.

Before You Go On

1. What is a pure substance?
2. What is matter made up of?
3. What are some examples of mixtures?

Lesson 1 – Review and Practice

Vocabulary

Complete each sentence with the correct word from the box.

mass	atoms	density
volume	properties	matter

1. Anything that takes up space and has mass is ______________.
2. About 5 million ______________ fit on the head of a pin.
3. You can measure ______________ with different kinds of containers.
4. You can measure ______________ with a balance.
5. Pennies have greater ______________ than puffed rice.
6. You can measure some ______________ of matter.

Check Your Understanding

Write the answer to each question.

1. What is matter? ______________________________
2. Name two ways you can learn about matter's properties. ______________________________
3. How do you measure the volume of something? ______________________________
4. A liter of water has more mass than a liter of air. Does water or air have greater density? Explain. ______________________________
5. What are atoms? ______________________________

Apply Science Skills

Science Reading Strategy: **Facts and Examples**

Reread page 52. Find a fact and an example. Complete the diagram.

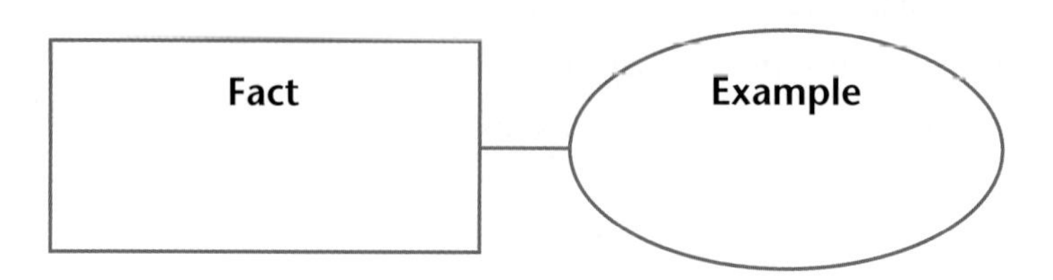

Using Visuals: **Micrographs**

The photograph on the left shows the substance graphite. Graphite is a pure substance. The micrograph on the right shows atoms of graphite.

Look at the photograph and the micrograph. Then answer the questions.

▲ Your pencil lead is made of graphite.

▲ Micrograph of graphite atoms

1. How are the photo and the micrograph different? How are they similar, or alike?

____

____

2. Graphite atoms are arranged in six-sided shapes. Can you see six-sided shapes in the micrograph? Point to them. _______________________

____

Discuss

What are some examples of matter that you see and use every day? What matter do you measure? How do you measure it?

For more practice, go to pages 65–66.

Practice Pages

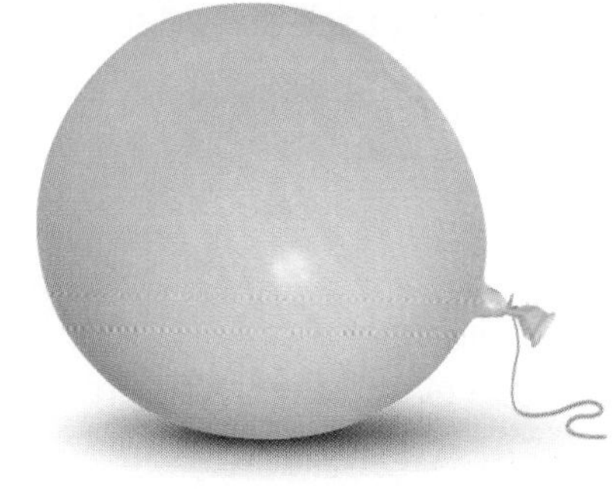

Name ______________________________ Date ______________

Before You Read

VOCABULARY

A. Match each key word with its definition. Write the letter.

_____ **1.** atoms	**a.**	how much of something there is
_____ **2.** matter	**b.**	the amount of mass in a given unit of volume
_____ **3.** mass	**c.**	very tiny things that make up all matter
_____ **4.** density	**d.**	anything that takes up space and has mass
_____ **5.** properties	**e.**	the amount of space something takes up
_____ **6.** volume	**f.**	traits or characteristic qualities

PRACTICE

B. Write four sentences using a key word and its definition.

1. ______________________________

2. ______________________________

3. ______________________________

4. ______________________________

C. Circle the correct word to complete each sentence.

1. Anything that takes up space and has mass is (matter / density).

2. You need a powerful microscope to see (mass / atoms).

3. You use scientific tools to determine the (properties / atoms) of something.

4. The amount of space that an object takes up is its (mass / volume).

5. (Mass / Volume) is like weight.

D. Write *T* for *true* or *F* for *false.* Correct the sentences that are false.

_____ **1.** Atoms are very large things that make up matter.

_____ **2.** A cup of sand has more density than a cup of feathers.

_____ **3.** You can use a measuring cup to measure mass.

_____ **4.** A material's properties help you know what the material is.

_____ **5.** Water takes up space and has mass.

Science Reading Strategy: Facts and Examples

A. Read the paragraph. Complete the diagram below with one more fact and an example for each.

All objects have mass, so all objects have density. But some objects are more dense than others. For example, suppose you throw a piece of wood in the water. It will float because wood has low density. Suppose you throw a piece of iron in the water. The iron will sink to the bottom because iron has high density.

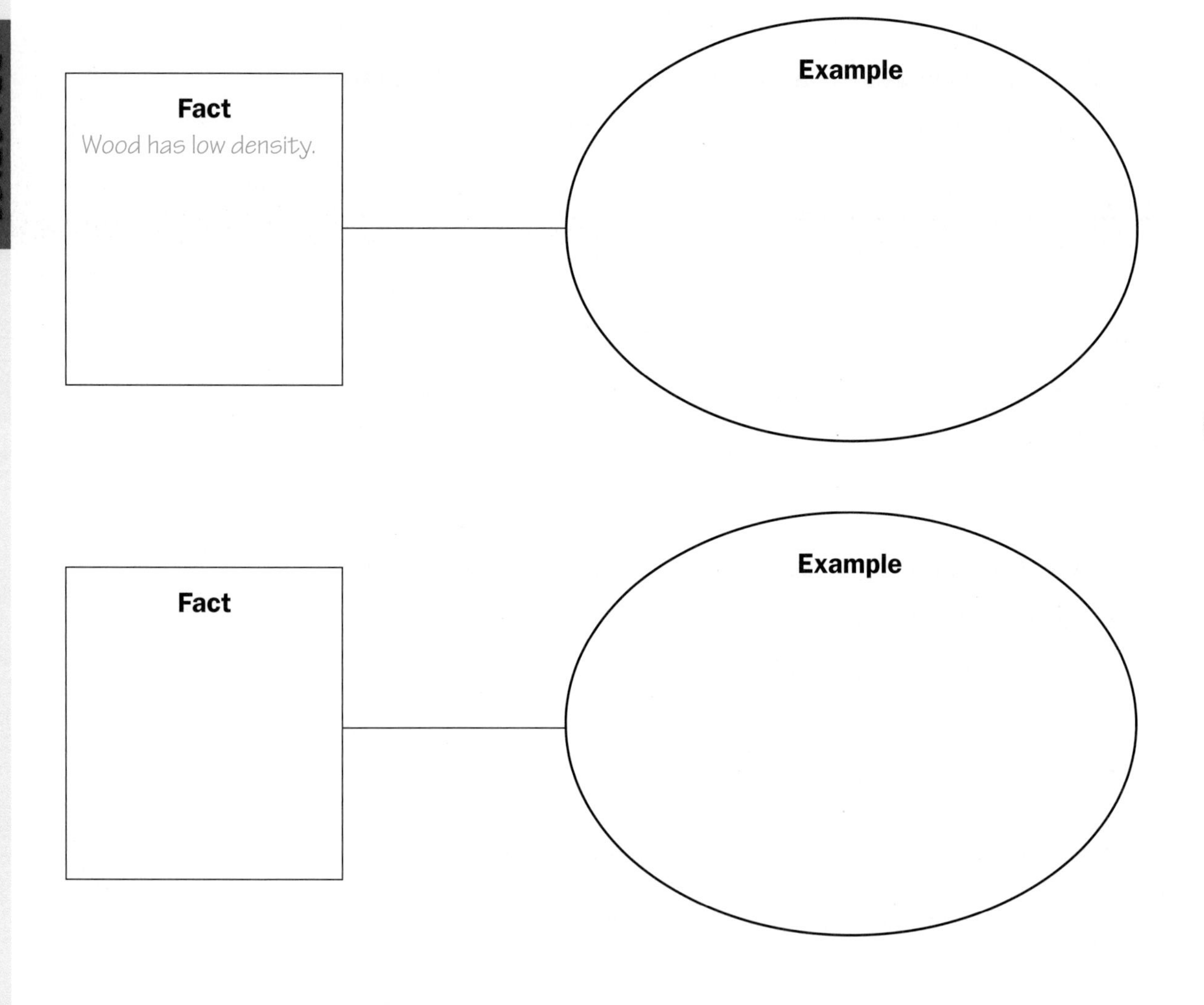

Name ______________________________ Date ______________

B. Read the paragraph. Complete the diagram below.

Everything around you is matter. Your desk, your classmates, and the air you breathe are all examples of matter. There are many different kinds of matter, but all matter is made up of atoms. Atoms are very tiny. Suppose you made a tiny dot with your pencil, no bigger than the period at the end of this sentence. Millions of atoms could fit on such a tiny dot. Each kind of matter has its own properties. For example, water is clear and has no color. It is a liquid. It freezes at 0° Celsius and boils at 100° Celsius.

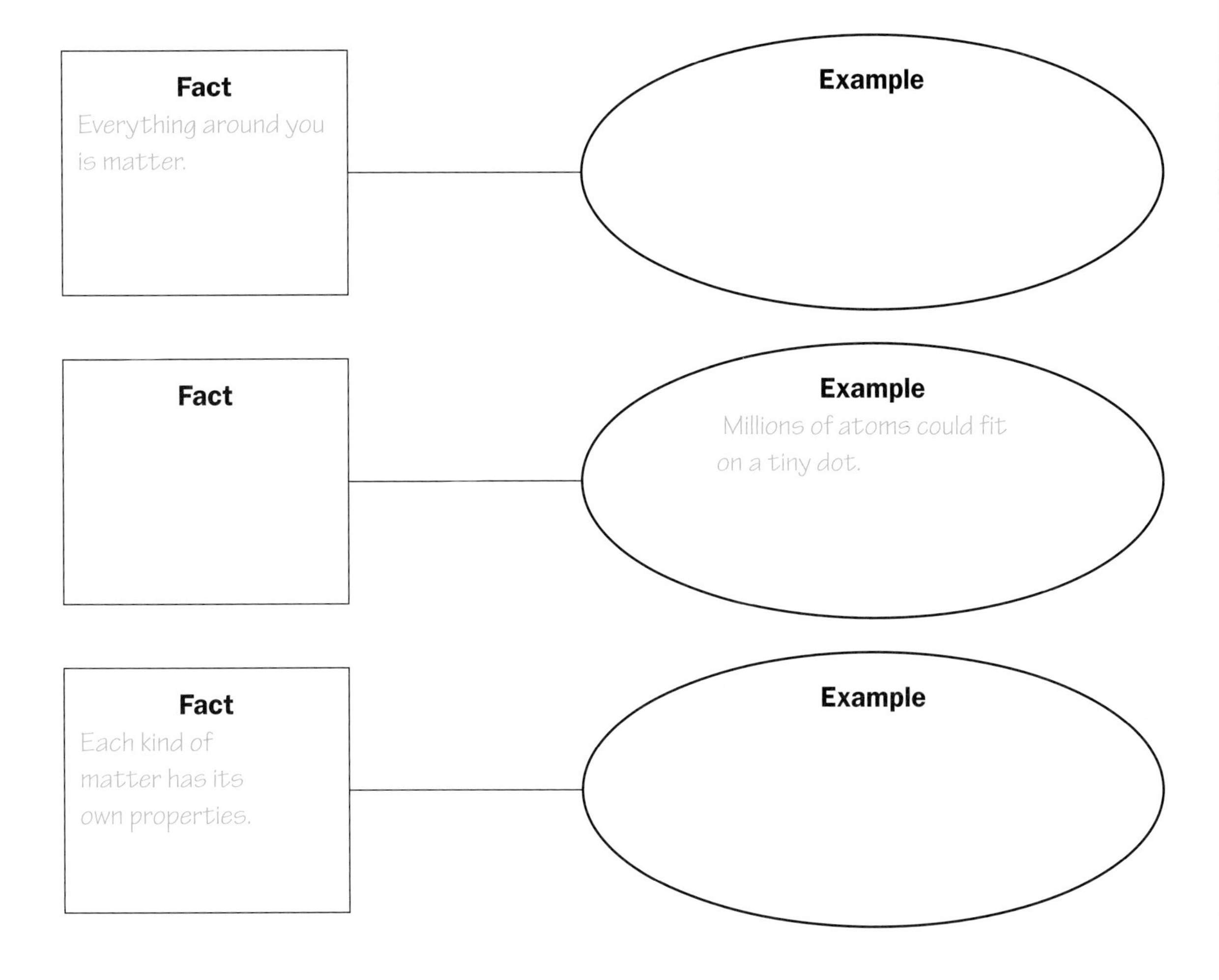

PRACTICE

Using Visuals: Micrographs

Look at the micrographs of sugar, a snowflake, salt, and sand. Then answer the questions.

▲ Sugar

▲ Snowflake

▲ Salt

▲ Sand

PRACTICE

1. Which substance is made of tiny rocks?

____

2. Which substances can you eat?

____

3. Which substance is a crystal with six branches?

____

4. Which micrograph was easiest to recognize?

____

5. Which is your favorite micrograph? Why?

____

Name ______________________________ Date ____________

Lesson 1 Review

VOCABULARY

Complete the puzzle. Use key words. Write the secret word.

1. _______ is like weight. __ __ (__) __
2. Everything around you is _______. __ __ __ __ (__) __
3. A cup of pennies has more _______ than a cup of popcorn. __ __ (__) __ __ __ __
4. _______ are tiny things. __ __ __ __ (__)
5. _______ can be measured in liters. __ __ __ __ __ (__)
6. We know what something is by looking at its _______. __ __ __ __ __ __ __ __ __ (__)

Secret word: __ __ __ __ __ __

VOCABULARY IN CONTEXT

Complete the paragraph. Use words from the box. There is one extra word.

matter	volume	properties	density	atoms	mass

All **(1)** _______________ is made of tiny things called **(2)** _______________. We can look at the matter's **(3)** _______________ (for example, its **(4)** _______________ and **(5)** _______________) to understand what it is.

CHECK YOUR UNDERSTANDING

Choose the best answer. Circle the letter.

1. Matter can be living or _______.
   - **a.** hard **b.** nonliving **c.** space
2. Color, shape, and size are _______ you can see.
   - **a.** balloons **b.** containers **c.** properties
3. A salad is a _______ of different kinds of matter.
   - **a.** mixture **b.** mass **c.** volume
4. _______ is an example of a pure substance.
   - **a.** Sand **b.** Salt **c.** Chocolate milk

PRACTICE

PRACTICE

APPLY SCIENCE SKILLS

Science Reading Strategy: Facts and Examples

Use the information in the paragraph to complete the diagram.

Each type of matter has its own properties. For example, honey is sweet and it is a liquid. Because honey is a liquid, it shares properties with other liquids. You can pour honey, and it takes the shape of its container.

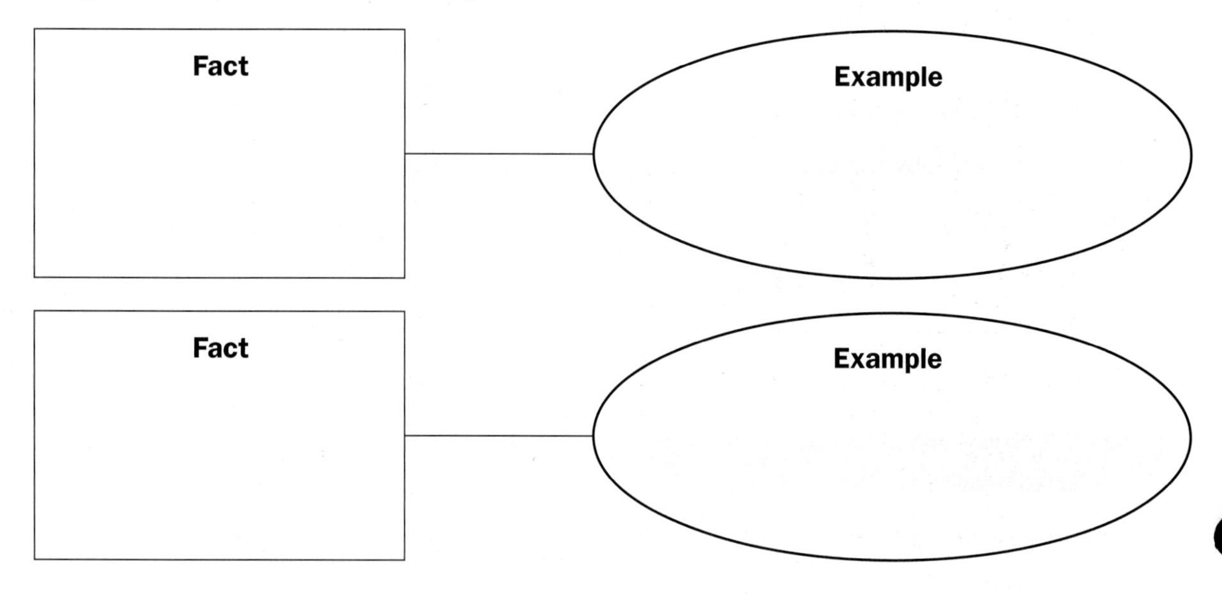

Using Visuals: Micrographs

Look at the micrographs of atoms on pages 47, 57, and 59. Then answer the questions.

1. Look at the micrograph on page 47. What substance do those atoms make?

____

2. Look at the micrograph on page 57. What substance do those atoms make?

____

3. Look at the micrograph on page 59. What substance do those atoms make?

____

4. Draw and label the shapes of the atoms in each micrograph.

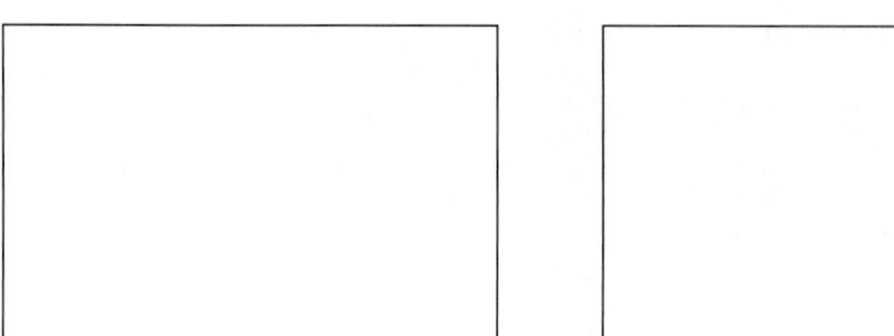
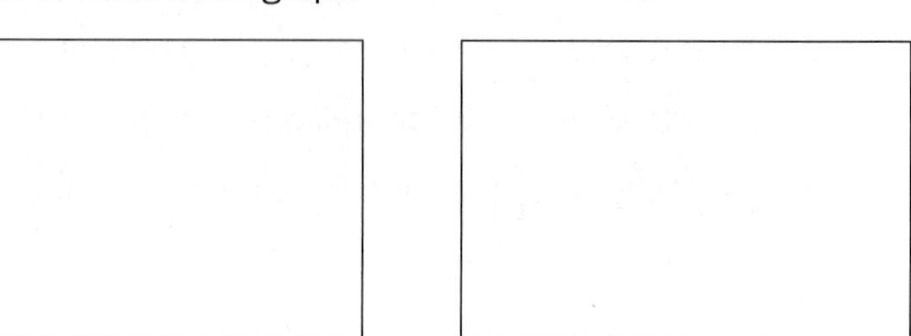

Science Journal

Write about five interesting things you have learned in this lesson.

1. ______________________________

____

____

2. ______________________________

____

____

3. ______________________________

____

____

4. ______________________________

____

____

5. ______________________________

____

____

Lesson 2 Before You Read

Vocabulary

▲ This photo shows three **states**, or forms, of matter. The glass container is a solid. The water is a liquid. The water vapor is a gas.

This water is boiling. The **boiling point** is the temperature at which a liquid changes to a gas. The boiling point of water is 100° Celsius (212° Fahrenheit). ▼

◀ These ice cubes are melting. The **melting point** is the temperature at which a solid melts. The melting point of water is 0° Celsius (32° Fahrenheit).

boiling point
chemical change
melting point
physical change
states

▲ Water made these chains rust. The groups of atoms that make up the chain changed into rust. This is an example of a **chemical change**.

▲ These cinnamon sticks have been ground into powder. It looks different, but it is still cinnamon. This is an example of a **physical change**.

Practice

Next to each key word, write its meaning. Use your own words.

boiling point *the temperature at which a liquid changes to a gas*

chemical change ______________________

melting point ______________________

physical change ______________________

states ______________________

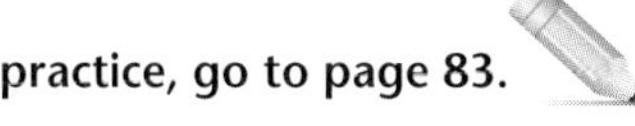

For more practice, go to page 83.

Science Skills

Science Reading Strategy: **Idea Maps**

As you read, draw an **idea map**. An idea map shows important ideas in the text. It helps you connect and understand the ideas you read.

Complete the idea map with ideas from page 51. You may need to add more circles.

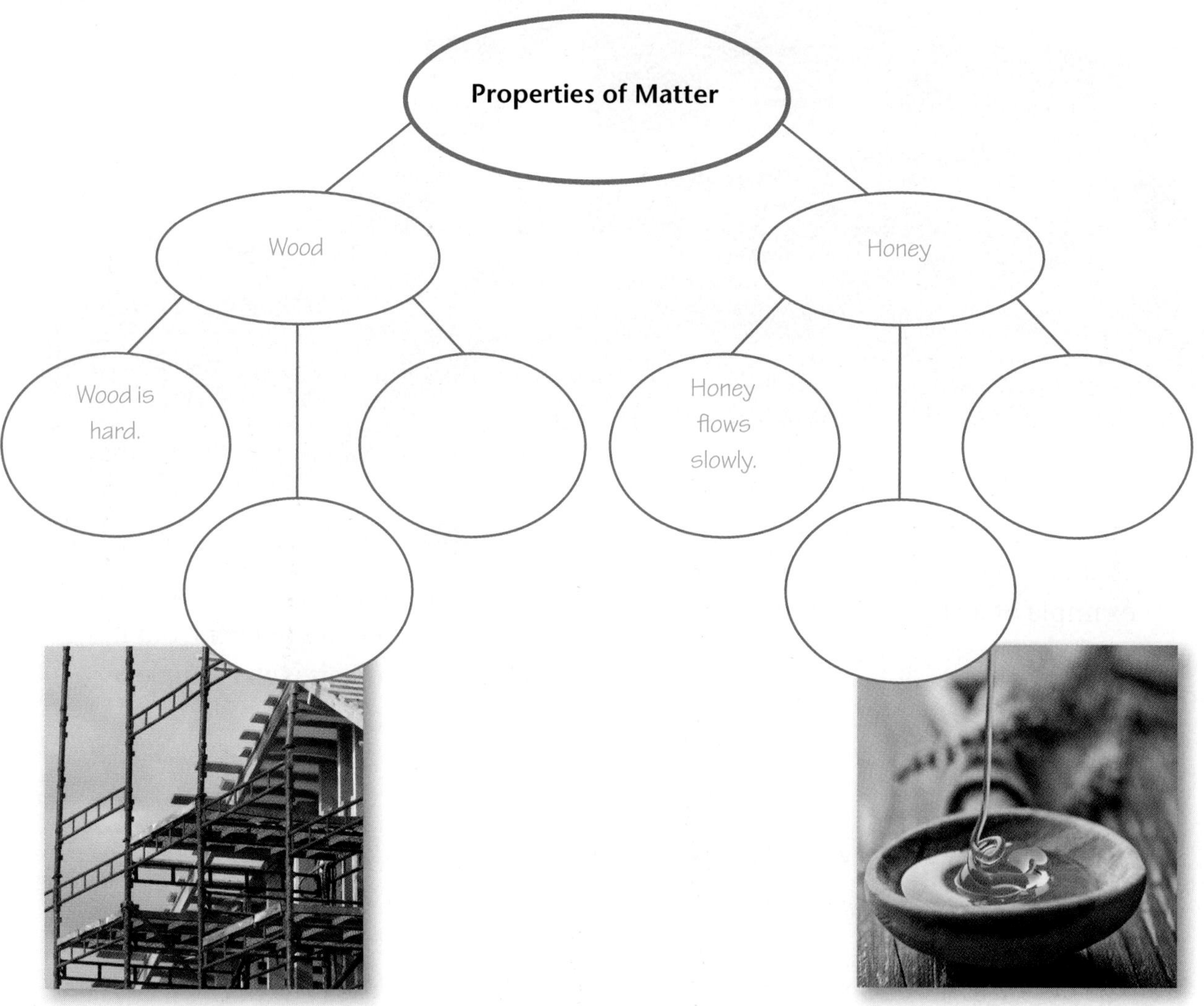

Look for an **As You Read** exercise in the lesson. It will ask you to complete an idea map. It will help you understand the lesson.

For more practice, go to pages 84–85. 

Using Visuals: **States of Matter Illustrations**

The illustrations below show what atoms do in each state of matter.
Look at the illustrations. Then answer the questions.

▲ Solid

▲ Liquid

▲ Gas

1. In which state do atoms move the most?
2. In which state do atoms move the least?
3. Do the atoms of a liquid move more or less than the atoms of a solid?

For more practice, go to page 86. 

States of Matter

> **LANGUAGE TIP**
> The word *state* has different meanings.
> *Ice is water in a solid* state.
> *Alaska is the largest* state *in America.*

Matter has three different states. This means that matter can have three different forms. Matter can be a solid, a liquid, or a gas.

Solids

Solid matter has a set shape. The shape of a solid stays the same. A solid also has a set volume. It takes up the same amount of space all the time.

▲ This kettle is a solid. It has a set shape and a set volume.

Liquids

Liquid matter flows. Water is an example of a liquid. Liquids do not have a set shape. The shape of a liquid can change. A liquid takes the shape of its container.

Liquids do have a set volume. Suppose you have 1 cup (about 250 milliliters) of water in a measuring cup. You then pour the water into a tall glass. The water takes the shape of the glass. But it still has the same volume: 1 cup.

▲ Each liquid takes the shape of its container.

SCIENCE AT HOME

Liquids, Shape, and Volume

Measure out 1 cup of a liquid. Pour the liquid into a container of any shape. Now pour the same liquid into a container with a different shape. What happens to the shape of the liquid? What happens to the volume of the liquid?

▼ Mercury is a liquid metal.

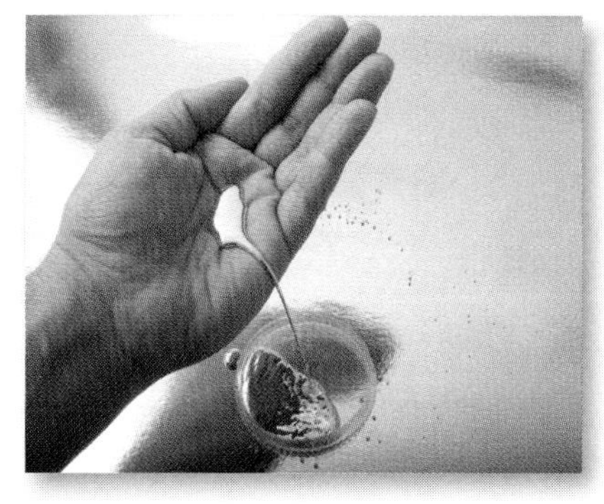

Before You Go On

1. Name the three different states of matter.
2. Do liquids have a set shape and volume?
3. What are the differences between a solid and a liquid?

Gases

Gases don't have a set shape. A gas takes the shape of its container. Gases don't have a set volume, either. A gas expands, or gets bigger, to fill its container. A small amount of gas expands to fill any container, no matter how large.

LANGUAGE TIP

don't → **do not**

The word *don't* is the same as do not.

As You Read

Idea Maps

Copy this idea map into your notebook.

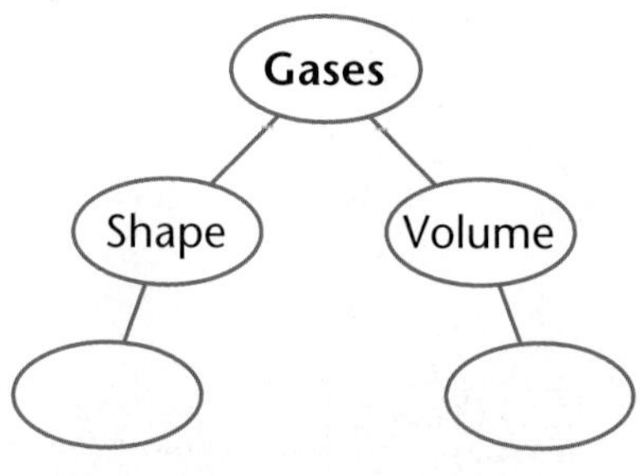

Complete the idea map with ideas on this page.

▲ When water changes into a gas, it becomes steam.

▲ This gas is expanding.

Changing States of Matter

Matter can change from one state to another. A solid can change to a liquid. A liquid can change to a gas.

The atoms in a solid are very close together. They do not move around. When a solid is heated, the atoms begin to move. With enough heat, the solid changes to a liquid.

The atoms in a liquid move around freely and are farther apart. When heated enough, the liquid turns into a gas. The atoms in a gas move fast and are far apart.

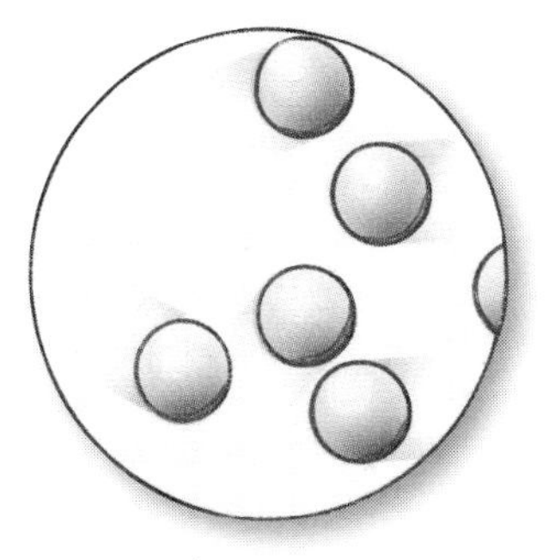

▲ Gas

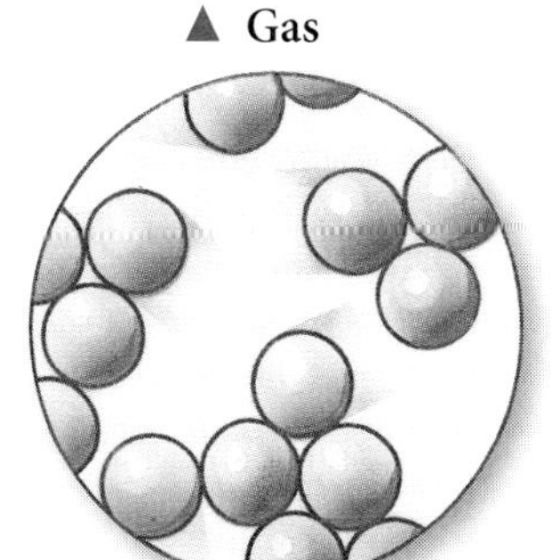

▲ Liquid

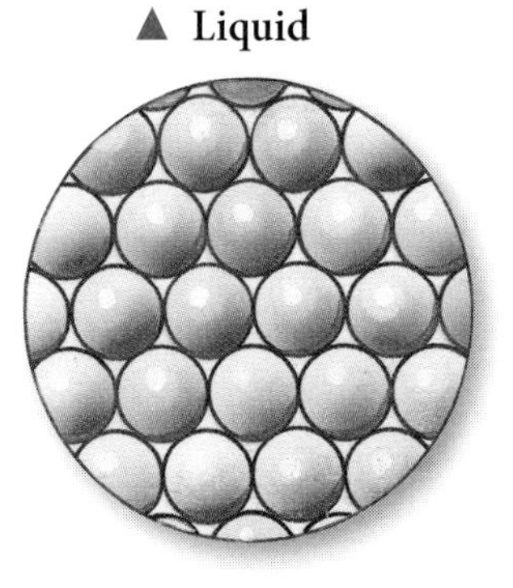

▲ Solid

▲ This photo shows water in all three states of matter. The solid (frozen) and liquid water are easy to see. There is also water vapor, water in its gas state, similar to a cloud, rising from the rushing water.

Melting Glaciers

Because of pollution, the Earth's temperature is rising. This causes solid glaciers to melt, or turn to liquid. The melting adds more water to the ocean. We have to control this melting. If we don't, some of the land we live on may soon be covered in water.

▲ This glacier in Norway is melting.

Before You Go On

1. Do gases have a set shape and volume?
2. What are the atoms in a solid like?
3. Can you think of an example of matter changing states?

The temperature at which a liquid changes to a gas is its boiling point. Each substance has its own boiling point. For example, water boils at 100° Celsius (212° Fahrenheit). Mercury boils at a much higher temperature—356.58° Celsius (674° Fahrenheit). A substance's boiling point is one of its properties.

▲ When water boils, it changes from a liquid to a gas.

SCIENCE NOW

Steam Power

When water boils, it turns into a gas often called steam. The steam can be used to make energy. This way of making energy is less expensive and safer than some other ways.

▲ This power station in New Zealand uses steam to make energy.

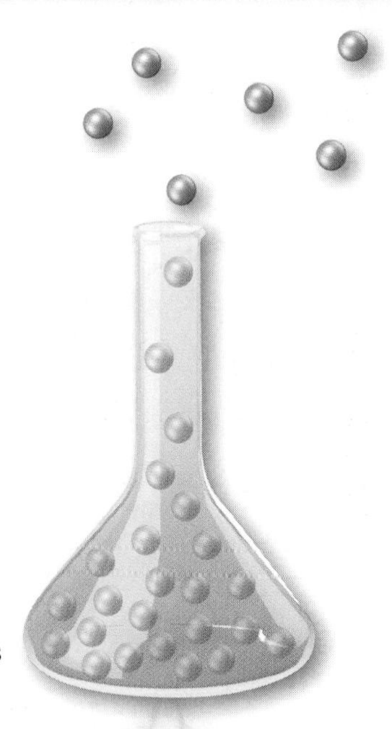

As a liquid boils, the gas expands and leaves the container. ▶

Melting point is the temperature at which a solid changes to a liquid. Each substance has its own melting point. For example, solid water (ice) melts at 0° Celsius (32° Fahrenheit). The melting point of copper is 1084.62° Celsius (1984.32° Fahrenheit). Like boiling point, a substance's melting point is one of its properties.

▲ When ice melts, water changes from a solid to a liquid.

Before You Go On

1. What is the boiling point of mercury?
2. What are two properties of a substance?
3. At what temperature do you need to store ice?

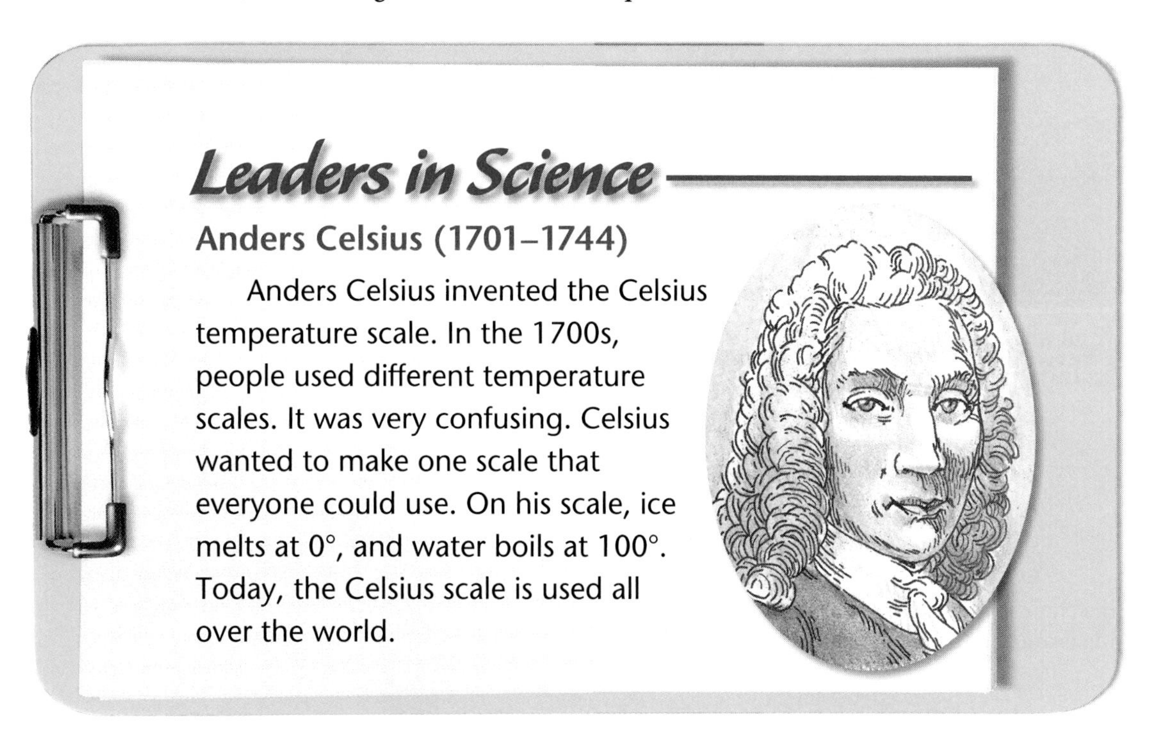

Leaders in Science

Anders Celsius (1701–1744)

Anders Celsius invented the Celsius temperature scale. In the 1700s, people used different temperature scales. It was very confusing. Celsius wanted to make one scale that everyone could use. On his scale, ice melts at 0°, and water boils at 100°. Today, the Celsius scale is used all over the world.

Physical and Chemical Changes

There are two main kinds of changes in matter: physical changes and chemical changes.

In a physical change, matter changes in appearance. But the groups of atoms that make up the matter do not change. One example of a physical change is changing state. When water boils, its groups of atoms do not change. It is still water, just in a different state.

Another example of a physical change is using wood to build a chair. The wood changes shape. But it is still wood.

LANGUAGE TIP

appear → verb
appearance → noun
After a physical change, matter appears *different. Its* appearance *changes.*

▲ This wood was cut from a tree. This is a physical change.

This chair is made out of wood. This is another physical change. ▶

During a chemical change, matter changes in appearance. The groups of atoms that make up the matter change as well. When this happens, a different kind of matter is formed. For example, iron turning to rust is a chemical change. The digestion of food is another chemical change.

The iron on this key is changing to rust. This is a chemical change. ▼

HISTORY CONNECTION

Stainless Steel

In 1912, a man named Harry Brearley invented a type of metal that does not rust. This metal is called stainless steel. People use stainless steel to make many products we use every day, such as knives.

Food goes through a chemical change inside your body. ▶

Before You Go On

1. What are the two main kinds of changes in matter?
2. What happens during a chemical change?
3. What are other examples of physical changes?

Lesson 2 – Review and Practice

Vocabulary

Complete each sentence with the correct word or phrase from the box.

boiling point	chemical change	states
physical change	melting point	

1. Solid, liquid, and gas are the three ____________ of matter.
2. A substance changes from a solid to a liquid at its ____________.
3. During a ____________, only the appearance of matter changes.
4. A substance changes from a liquid to a gas at its ____________.
5. During a ____________, the groups of atoms that make up matter change.

Check Your Understanding

Write the answer to each question.

1. How are solids different from gases? ____________________

____

2. Describe the atoms in a liquid. ____________________

____

3. What's an example of a physical change? ____________________

____

4. What's an example of a chemical change? ____________________

____

Apply Science Skills

Science Reading Strategy: **Idea Maps**

Complete the map with ideas you learned on pages 78–79.

Changes in Matter

Physical Changes

Chemical Changes

Using Visuals: **States of Matter Illustrations**

This graph and the illustrations show water at different temperatures. The line on the left shows the temperature in Celsius rising from cold to hot. The line at the bottom shows the movement of the atoms. The graph also shows the melting point and the boiling point of water.

Look at the graph and the illustrations. Then answer the questions.

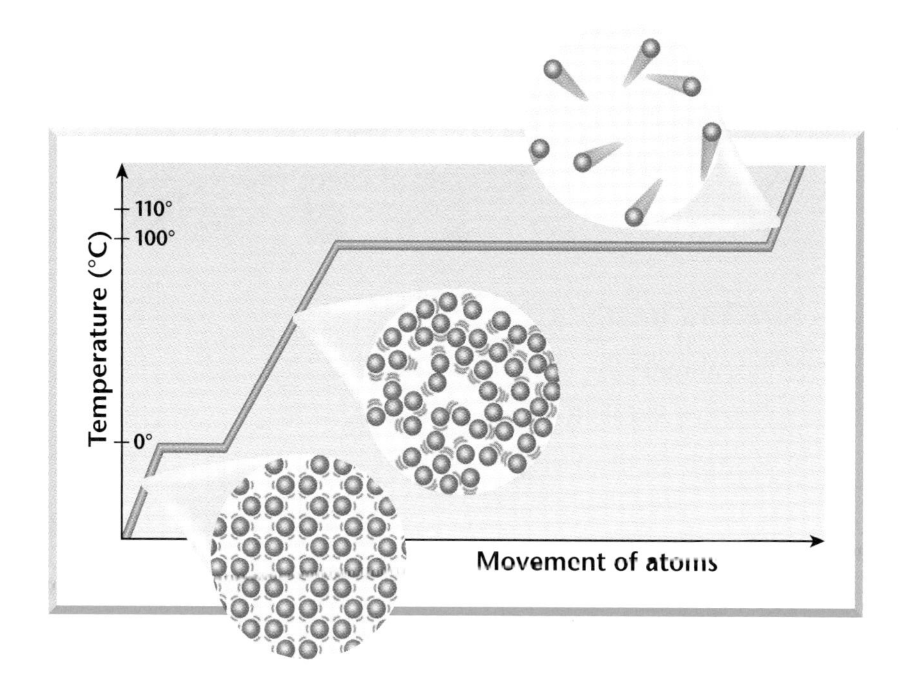

1. What happens to the atoms as the temperature rises from below 0°C to 80°C?

____

2. What happens to the atoms as the temperature rises from 80°C to over 100°C?

____

Discuss

Many glass, metal, and plastic items are made by changing the substance's state. A solid is melted. Then the liquid is cooled and shaped to make another solid. What things do you see around you that were made this way?

For more practice, go to pages 87–88. 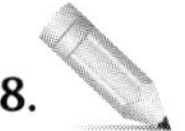

Practice Pages

PRACTICE

Name ______________________ Date ____________

Before You Read

VOCABULARY

A. Match each key word with its definition. Write the letter.

_____	**1.** physical change	**a.** forms of matter: solid, liquid, and gas
_____	**2.** chemical change	**b.** the temperature at which a solid melts
_____	**3.** boiling point	**c.** the groups of atoms in the matter do not change
_____	**4.** states	**d.** the groups of atoms in the matter change
_____	**5.** melting point	**e.** the temperature at which a liquid changes to a gas

B. Write five sentences using each key word and its definition.

1. ______________________

2. ______________________

3. ______________________

4. ______________________

5. ______________________

C. Circle the correct word to complete each sentence.

1. The three (states / points) of matter are solid, liquid, and gas.

2. A solid changes to a liquid at its (boiling / melting) point.

3. A liquid changes to a gas at its (boiling / melting) point.

4. Mashing corn is an example of a (physical / chemical) change.

5. Rusting metal is an example of a (physical / chemical) change.

D. Write *T* for *true* or *F* for *false.* Correct the sentences that are false.

_____ **1.** There are three states of matter: solid, liquid, and space.

_____ **2.** Water vapor is a gas.

_____ **3.** The boiling point of water is 0° Celsius.

_____ **4.** When ice melts, water changes from a solid to a liquid.

_____ **5.** Water changing states is a chemical change.

PRACTICE

Science Reading Strategy: Idea Maps

A. Complete the idea map below with ideas you learned in Lesson 1.

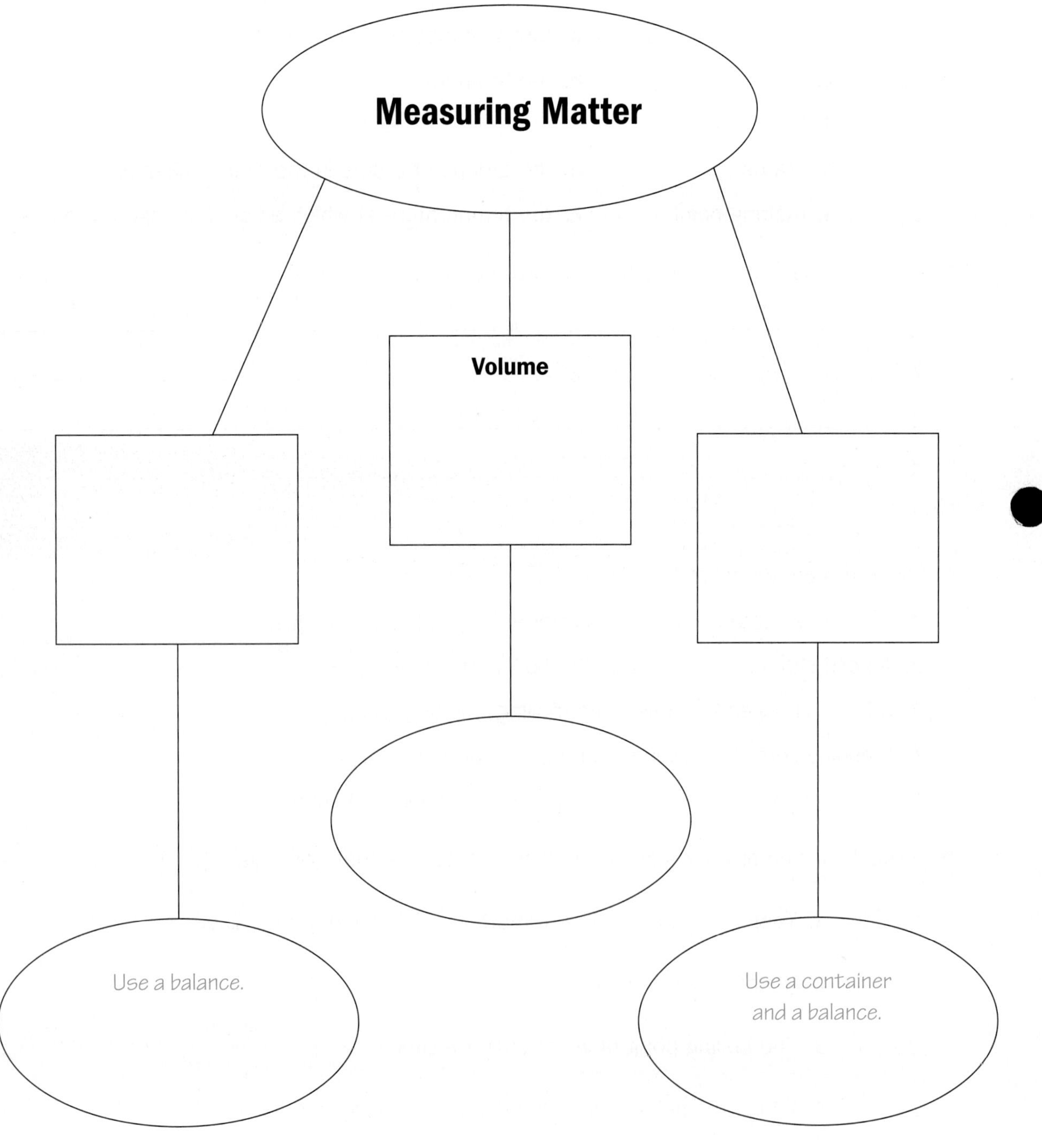

Name ______________________________ Date ______________

B. Complete the idea map below with two kinds of matter for each of the three states.

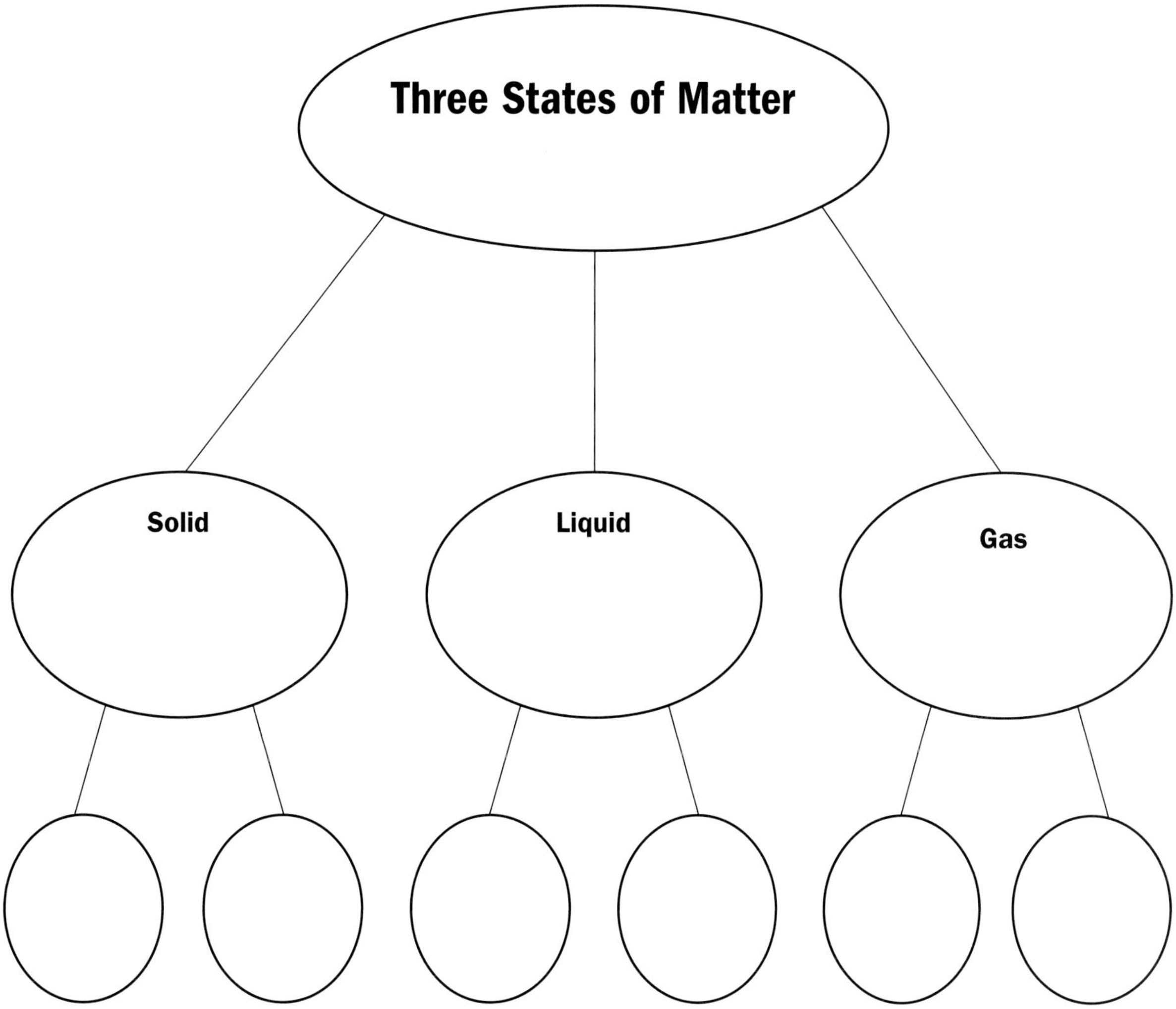

Three States of Matter

Solid

Liquid

Gas

PRACTICE

Using Visuals: States of Matter Illustrations

Look at the illustrations below. Then answer the questions.

_______________ _______________ _______________

1. Which state of matter does each illustration show? Label the illustrations.

____

2. What does each round shape represent?

____

3. In which state of matter is there the most space between atoms?

____

4. In which state is there the least space between atoms?

____

5. How would you change a liquid into a gas?

____

PRACTICE

Name ______________________________ Date ______________

Lesson 2 Review

VOCABULARY

Use words from the box to identify each clue. Write the word or phrase on the line.

chemical change	melting point	physical change	states	boiling point

1. I am the temperature at which a liquid changes to a gas. ______________
2. I change groups of atoms to make a new type of matter. ______________
3. I am the temperature at which a solid changes to a liquid. ______________
4. I change only the way matter looks. ______________
5. I am also called forms of matter. ______________

VOCABULARY IN CONTEXT

Complete the paragraph. Use words from the box.

boiling point	chemical change	states	melting point	physical change

There are three **(1)** ______________ of matter: solid, liquid, and gas. These three forms of matter can change. For example, at the **(2)** ______________, ice changes to liquid. At the **(3)** ______________, water changes to vapor, or steam. Matter can change in two main ways. Corn can be mashed. That is a **(4)** ______________. A metal chain can rust. That is a **(5)** ______________ .

CHECK YOUR UNDERSTANDING

Choose the best answer. Circle the letter.

1. The atoms in a ______ move around freely.
   - **a.** solid **b.** liquid **c.** chemical
2. A rise in ______ causes atoms to move faster.
   - **a.** temperature **b.** density **c.** mass
3. The digestion of food is an example of a ______.
   - **a.** physical change **b.** boiling point **c.** chemical change

PRACTICE

APPLY SCIENCE SKILLS

Science Reading Strategy: Idea Maps

Complete the idea map with ideas from this lesson.

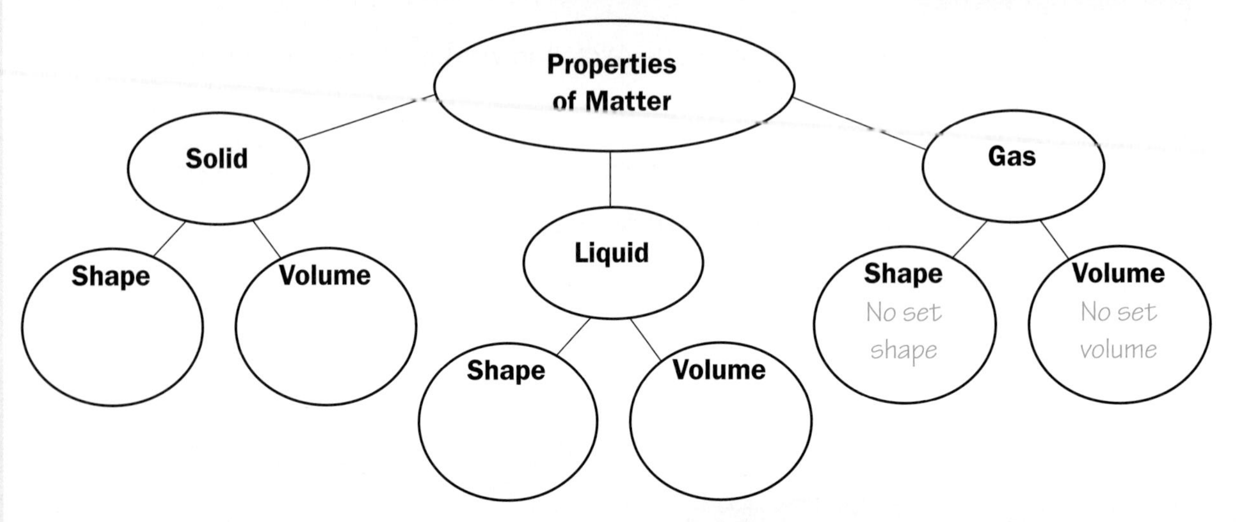

Using Visuals: States of Matter Illustrations

Look at the illustrations below. Then answer the questions.

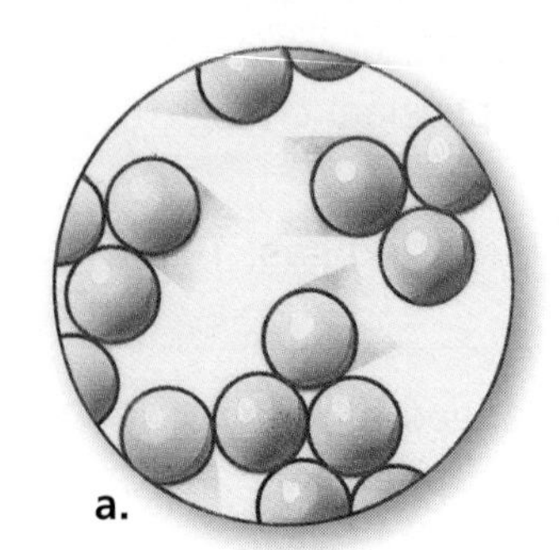

a.

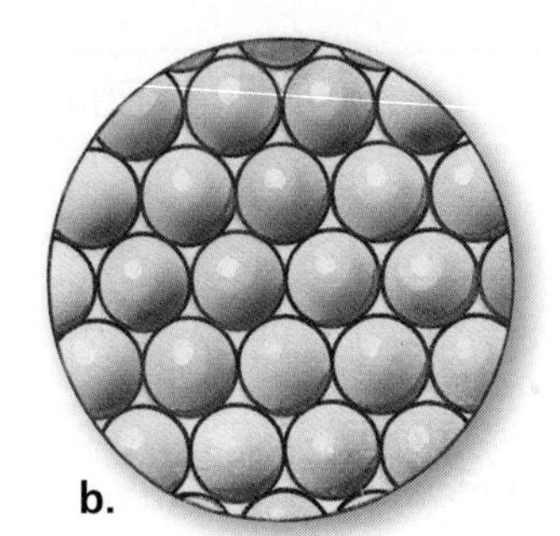

b.

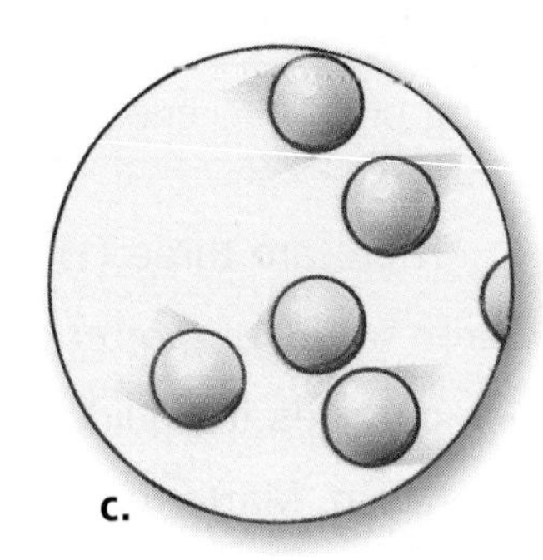

c.

1. What is illustration a. an example of?

____

2. What is illustration b. an example of?

____

3. What is illustration c. an example of?

____

4. In which state do the atoms move the fastest?

____

5. In which state do the atoms move the slowest?

____

PRACTICE

Science Journal

Write about five interesting things you have learned in this lesson.

1. ______________________________
____
____

2. ______________________________
____
____

3. ______________________________
____
____

4. ______________________________
____
____

5. ______________________________
____
____

Part Review

Vocabulary

Answer the questions with words or phrases from the box.

mass	melting point	volume	measure
physical change	matter	atoms	chemical change
density	states	boiling point	

1. What property can you measure with a container? ______
2. What property can you measure with a balance? ______
3. How do you find out the mass of something? ______
4. What is anything that takes up space and has mass? ______
5. What tiny pieces make up matter? ______
6. What is the mass of something per unit of volume? ______
7. What type of change happens when matter changes state? ______
8. What type of change happens to food when you eat it? ______
9. What is another word for the three forms matter can have? ______
10. At what temperature does a substance change from a solid to a liquid? ______
11. At what temperature does a substance change from a liquid to a gas? ______

Check Your Understanding

Write *T* for *true* or *F* for *false*. Then rewrite each false statement to make it correct.

______ 1. Salt is a mixture. ______

______ 2. Rust is an example of a chemical change. ______

______

______ 3. Ice melting is an example of a chemical change. ______

______

______ 4. Liquids have a set shape. ______

______ 5. You cannot use properties to tell substances apart. ______

______

Extension Project

With a partner, make a chart with three columns. Label the columns *Solids*, *Liquids*, and *Gases*. Complete the chart with different kinds of matter. Write each word in the correct column.

Apply Science Skills

Using Visuals: **States of Matter Illustrations**

Look at the numbered illustrations. They show the different states of water. Then look at the sentences on the left. Next to each, write the number of the illustration it describes.

_______ **1.** The temperature is higher than 100°C.

_______ **2.** The temperature is higher than 0°C.

_______ **3.** The temperature is lower than 0°C.

_______ **4.** It has a set shape and volume.

_______ **5.** It has a set volume but not shape.

_______ **6.** It has no set shape or volume.

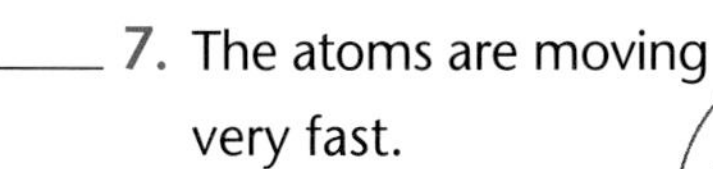

_______ **7.** The atoms are moving very fast.

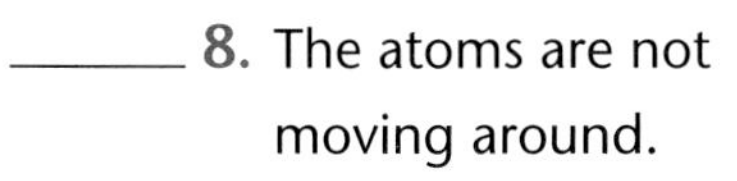

_______ **8.** The atoms are not moving around.

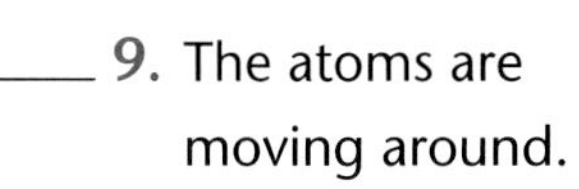

_______ **9.** The atoms are moving around.

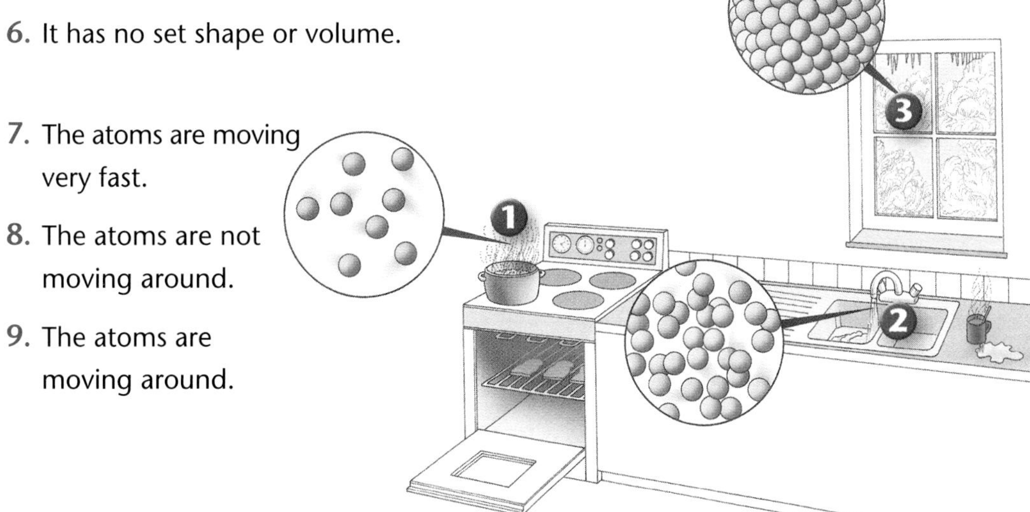

Part Experiment

Can You Observe Density?

You can see density. When you mix different liquids together, they can separate. They can form layers based on their densities. Any solids added will float, sink, or do neither based on their densities.

Purpose

To observe the density of different liquids and solids

Materials

large clear jar
water
oil
honey or syrup
solids (piece of cork, small plastic toy, coin, grape or cherry)

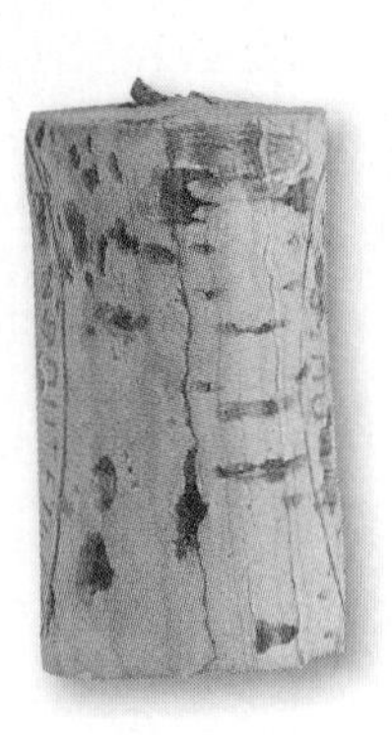
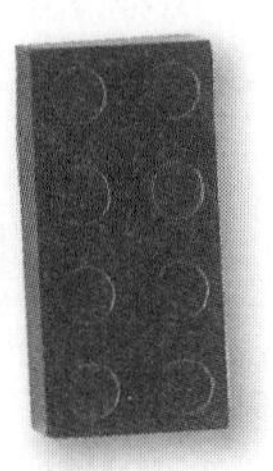

What to Do

1. Measure out 1 cup of oil, 1 cup of water, and 1 cup of honey or syrup.
2. Add the liquids to the jar one at a time. The order does not matter.
3. Let the liquids sit for a few minutes. Observe them.
4. Add the solids one at a time.
5. Wait a few minutes. Then observe the solids.

Draw Conclusions

1. What happened to the liquids? What does this tell you about their densities?
2. What happened to the solids? What does this tell you about their densities compared to the densities of the liquids?

Work with a partner. Discuss your conclusions. Then write them under Step 5 on page 94.

Experiment Log: Can You Observe Density?

Follow the steps of the scientific method as you do your experiment. Write notes about each step as the experiment progresses.

Step 1: Ask questions.

Step 2: Make a hypothesis.

Step 3: Test your hypothesis.

Step 4: Observe.

Step 5: Draw conclusions.

Write About It

1. Physical and chemical changes are happening around you all the time. Write about physical and chemical changes you have personally observed. What happened in each change?

2. Write about the foods you eat. Which foods are mixtures? Which foods can change from one state to another? Give examples.

Sound and Light

Part 2

Part Concepts

Lesson 1

- Vibrations make sounds.
- Sound waves move out in all directions.
- Sound waves bounce off hard surfaces.
- Sound has high and low pitches.
- Volume is the loudness or softness of sound.
- Sound waves move at different speeds through solids, liquids, and gases.

Lesson 2

- Light is an electromagnetic wave.
- Light bounces off surfaces as reflections.
- Light passes through some matter.
- Light can bend.
- There are different colors of light.
- The color of light depends on its wavelength and frequency.

Get Ready

Listen. What sounds do you hear? People talking? A bell ringing? Are the sounds soft or loud? Write the sounds you hear in the chart. Next to each sound, write the word *soft* or *loud*.

Sound	Soft or Loud?

Vocabulary

▲ This cymbal is vibrating, or moving up and down. The **vibrations** make a sound.

Vibrations move through the air in waves. Our ears can hear these **sound waves.** ▼

▲ These bass trumpets are making the air vibrate slowly. They are making low-**frequency** sound waves. The sound the bass trumpets make has a deep, low **pitch**.

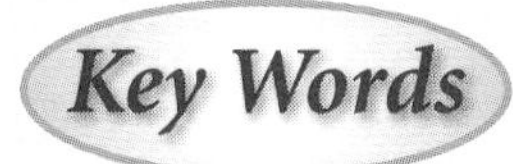

Key Words

echo
frequency
pitch
sound waves
vibrations
volume

▲ **Volume** is how soft or loud a sound is. A shout is a high-volume sound. Shout near a hard surface, and you can hear an **echo**. The sound waves bounce off the hard surface. You hear the sound again.

Practice

Choose the word that completes each sentence.

1. A low-frequency sound wave has a low ________.
   a. pitch **b.** volume **c.** echo
2. You hear an ________ when sound bounces off a hard surface.
   a. volume **b.** echo **c.** pitch
3. Something vibrating slowly makes a sound wave with a low ________.
   a. frequency **b.** volume **c.** echo
4. ________ is how soft or loud a sound is.
   a. Pitch **b.** Frequency **c.** Volume
5. When something moves up and down, it makes ________.
   a. vibrations **b.** pitch **c.** volume
6. Vibrations move through the air in ________.
   a. volume **b.** frequency **c.** sound waves

For more practice, go to page 113. 

Science Skills

Science Reading Strategy: **Act It Out**

When you read new information, it's helpful to **act it out.** Use your body and objects around you to act out the information. This will help you understand what you read.

Read this text. Act out the information in each paragraph.

> We make sounds with the vocal cords in our throats. We push air from our lungs into our throats. Then we make our vocal cords vibrate. This makes a sound. You can put your fingers on your throat to feel this vibration.
>
> When we make our vocal cords tight, we make a sound with a high pitch. When our vocal cords are loose and relaxed, we make a sound with a low pitch.
>
> Pushing more air past our vocal cords makes a louder sound. Pushing less air past our vocal cords makes a softer sound.

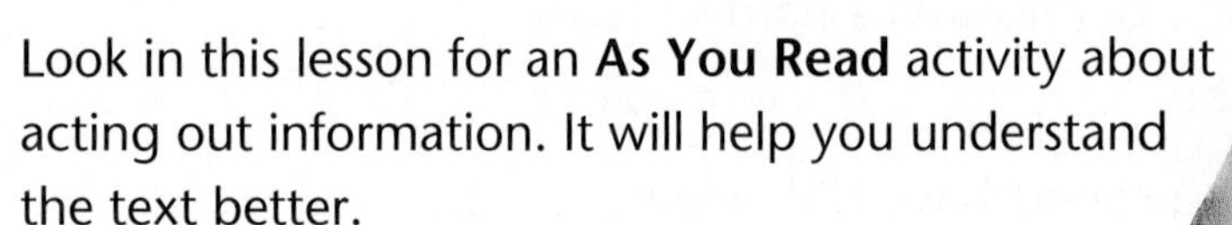

Look in this lesson for an **As You Read** activity about acting out information. It will help you understand the text better.

This girl touches her throat as she makes a sound. She feels her vocal cords vibrate. ▶

For more practice, go to page 114.

Using Visuals: **Charts**

The chart below shows the volume of different sounds. Volume is how soft or loud a sound is. We measure volume in units called **decibels**. The numbers in the chart show decibels.

Look at the chart. Then discuss the questions with a partner.

▼ Rain

▼ Plane

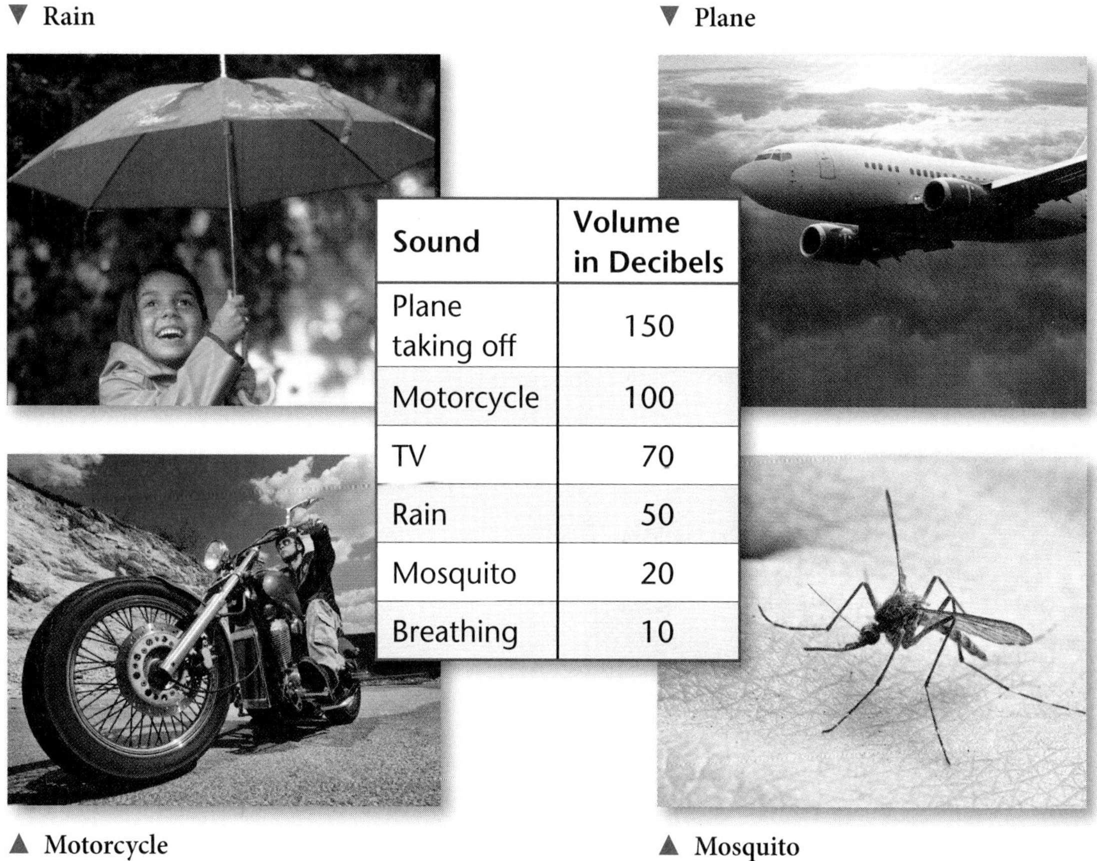

Sound	Volume in Decibels
Plane taking off	150
Motorcycle	100
TV	70
Rain	50
Mosquito	20
Breathing	10

▲ Motorcycle

▲ Mosquito

1. Which sound is the softest? Which is the loudest?
2. Which is louder, rain or a mosquito?
3. By how many decibels is a plane taking off louder than a motorcycle?

decibels: units used to measure volume

For more practice, go to pages 115–116.

Lesson 1

Sound

Sound is very important in our lives. We use sound to communicate with each other. We listen and dance to the sounds we call music. We look around when we hear a loud sound. This helps keep us safe. But what is sound?

▲ Hitting the surface of a drum makes it vibrate. The drum's vibrations make air vibrate, too.

Hit a drum. What happens to the surface of the drum? It vibrates, or moves up and down. This vibration makes a sound.

A vibration causes the matter around it to vibrate, too. A drum's vibration makes the atoms in the air vibrate. The vibration moves through the air and reaches your ears. It makes your eardrums vibrate. You hear a sound.

LANGUAGE TIP

vibrate → verb

vibration → noun

▲ This girl is listening to music with headphones. The speakers in the headphones are vibrating.

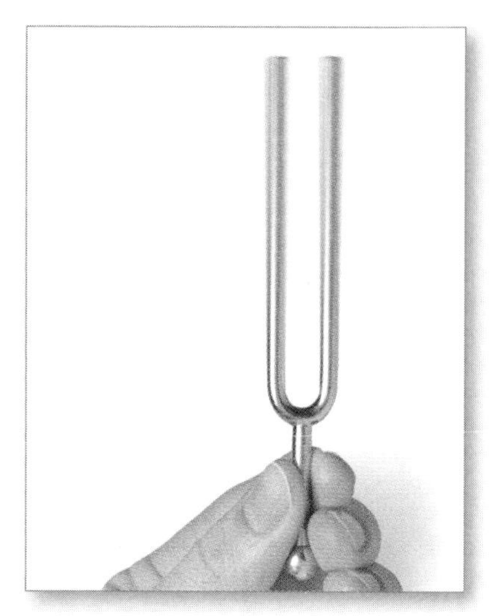

▲ This tuning fork is vibrating.

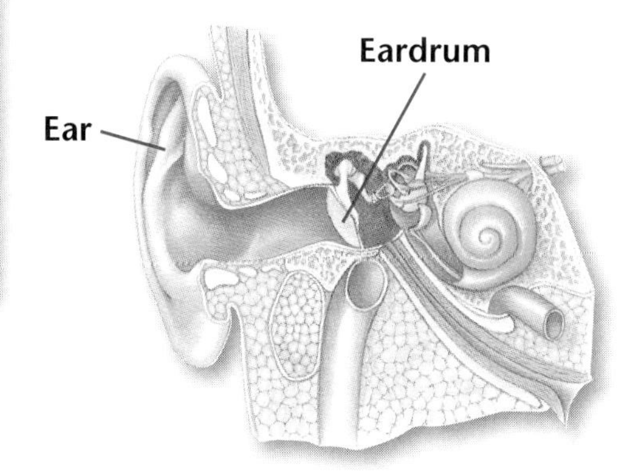

▲ Our eardrums catch vibrations.

Before You Go On

1. What makes sounds?
2. What part of your ear catches vibrations?
3. How does hearing help keep you safe?

Sound Waves

Throw a rock into water. The rock makes waves that move out in circles. Vibrations move in sound waves. Sound waves move out in circles, too. But they move out in all directions. They move up, down, and to the left and right. That's why you can hear sounds coming from above, below, and all around you.

▲ The waves in this water are moving out in circles.

POETRY CONNECTION

Basho

Basho was a poet who lived in Japan in the 1600s. He wrote many poems. Here is one of his most famous:

A frog jumps in an old pond.
The sound of water!

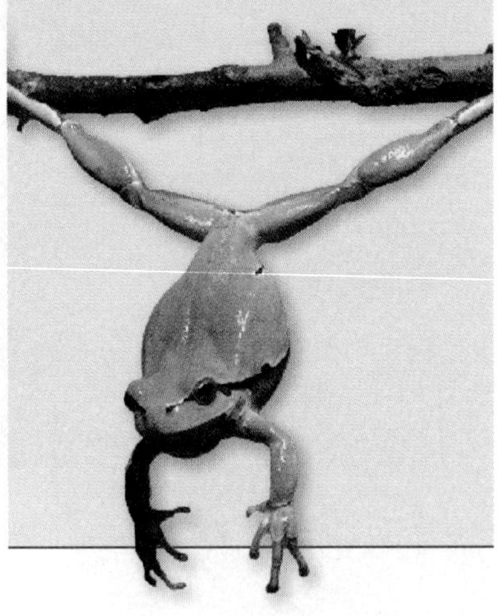

▲ The sound waves from a shout go out in all directions.

Echoes

Sound waves hit hard surfaces and bounce off. Shout in a place with a lot of hard surfaces, like a parking garage. The sound waves will hit the hard surfaces and bounce off. You can hear an echo.

Shout in a room with a lot of soft things like cushions, furniture, and carpet. Soft objects **absorb** a lot of sound waves. The sound waves don't bounce around much. It's hard to hear an echo.

absorb: take in and hold, like a sponge takes in and holds water

SCIENCE AT HOME

Making Echoes

You can make echoes in your home. Stand in a room without furniture or carpet, like a bathroom or hall. Say "Hi!" very loudly. Listen for the echo. Then stand in a room with furniture and carpet. Say "Hi!" very loudly again. Do you hear an echo? If yes, how is it different?

▲ This parking garage has a lot of hard surfaces. You can hear an echo here.

▲ This room has carpet and soft furniture with cushions. It's hard to hear an echo here.

Before You Go On

1. How do sound waves move?
2. Where can you hear echoes?
3. Where's a good place in your school to hear an echo?

Pitch and Frequency

Pitch is how high or low something sounds to our ears. A whistle has a high pitch. A double bass has a deep, low pitch. The pitch we hear is caused by the frequency of the sound waves. A whistle makes the air vibrate quickly. The sound waves are close together. They have a high frequency, and our ears hear a high pitch. A double bass makes the air vibrate more slowly. The sound waves are farther apart. They have a lower frequency, and our ears hear a lower pitch.

LANGUAGE TIP

The adjective *frequent* means "happening often." The noun form of *frequent* is *frequency*.

▲ This whistle has a high pitch.

▲ This double bass has a low pitch.

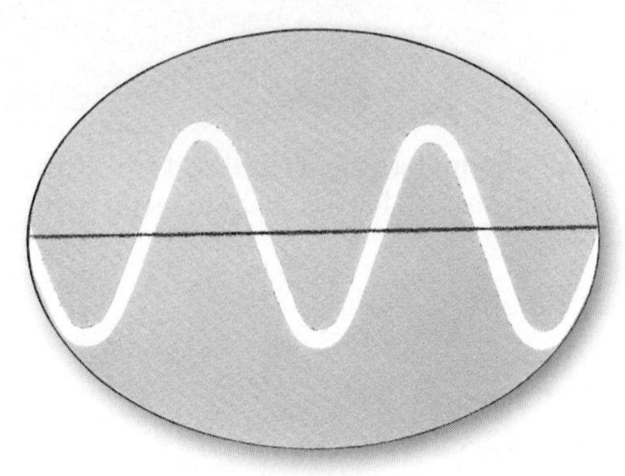

▲ High-frequency waves

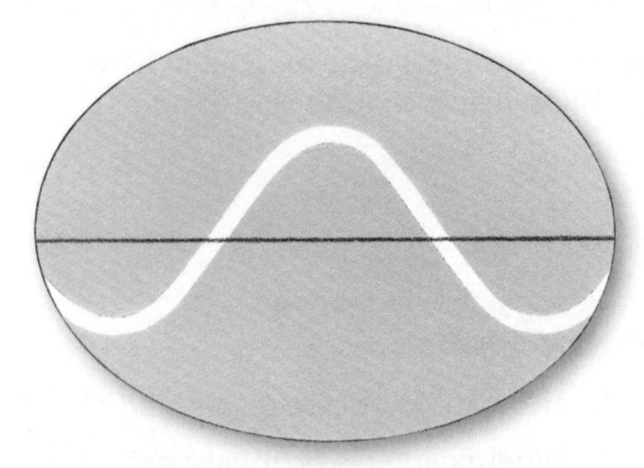

▲ Low-frequency waves

Exploring

Animals and Frequencies

Scientists measure the frequency of sound waves in units called hertz (Hz). Different animals can hear sounds of different frequencies.

◀ The lowest-frequency sound humans can hear is 20 Hz. The highest frequency we can hear is 23,000 Hz.

Elephants can hear sounds of 16 to 12,000 Hz. They can't hear some of the high pitches we hear. But they can hear lower pitches than we can. ▶

◀ Mice can hear sounds of 1,000 to 91,000 Hz. They can hear higher pitches than we can. Mice communicate using these high-frequency sounds.

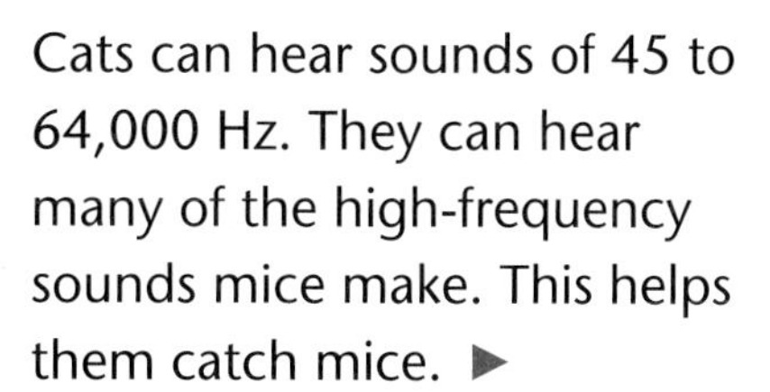

Cats can hear sounds of 45 to 64,000 Hz. They can hear many of the high-frequency sounds mice make. This helps them catch mice. ▶

Before You Go On

1. What makes a sound have a particular pitch?
2. Are low-frequency sound waves close together or far apart?
3. Do you think people can hear all the sounds mice make? Why or why not?

Volume

Volume is the loudness or softness of sound. A shout is loud. It has a high volume. It takes a lot of energy to make this sound. And the sound has a lot of energy. Tap on your desktop with your finger. When you hit softly, the sound is soft. The sound doesn't have much energy, so its volume is low. When you tap harder, and use more energy, the sound is louder. The sound has more energy, so its volume is higher.

We can measure the volume of sound in decibels. High-decibel sounds can hurt our ears.

Act It Out
As you read, act out the information in the first paragraph.

LANGUAGE TIP

The word *volume* has more than one meaning. *Volume* is the amount of space something takes up. *Volume* is also the softness or loudness of sound.

This space shuttle makes a very loud sound when it takes off. The sound has a lot of energy. ▶

Sound	Volume in Decibels
Shuttle taking off	190
Rock concert	115
Busy street	70
Classroom or office	45
Falling leaves	10

Volume in decibels ▶

Sound and Matter

Sound waves move only through matter (solids, liquids, or gases). Shout on the moon, and you will not hear a sound. The moon doesn't have an atmosphere. There are no gases for the sound waves to travel through.

Sound waves move most quickly through solids. The atoms in solids are close together. The vibration moves quickly from atom to atom. Sound moves more slowly through liquids. The atoms are farther apart. It takes longer for the vibration to reach each atom. Sound moves slowest through gases because the atoms are far apart. It takes a lot of time for the vibration to move from atom to atom.

▲ **You can't make sounds on the moon.**

▲ **Sound travels faster through water than through air.**

Before You Go On

1. Which sounds have more energy, soft or loud sounds?
2. Does sound move fastest through solids, liquids, or gases? Why?
3. Are there sounds on Venus? Why or why not?

Lesson 1 –Review and Practice

Vocabulary

Complete each sentence with the correct word or phrase from the box.

pitch	vibrations	frequency
volume	echo	sound waves

1. You hear an ____________________ when sound waves bounce off a hard surface.
2. ____________________ make sound waves.
3. A double bass makes low-____________________ sound waves and has a low pitch.
4. A loud sound has a high ____________________.
5. The sound of a whistle has a high ____________________.
6. Vibrations move through matter in ____________________.

Check Your Understanding

Write the answer to each question.

1. How do people hear sounds? ____________________
2. What directions do sound waves move in? ____________________
3. Which sounds have higher volume, loud or soft sounds? ____________________
4. What kind of matter does sound travel through fastest? ____________________

Apply Science Skills

Science Reading Strategy: **Act It Out**

Reread the last paragraph on page 109. Act it out. Get a pen, ruler, or other hard object. Tap it on your desk. Listen to the sound. Then put your ear to the desk and tap again. Can you hear sound traveling through your desk? Are the sounds different? Explain.

____

____

____

Using Visuals: **Charts**

The chart below shows the speed of sound in different types of matter. It shows the speed in both meters per second and feet per second.

Look at the chart. Then answer the questions.

Matter	Speed of Sound	
	Meters per second	Feet per second
Dry, cold air	343	1,125
Water	1,550	5,085
Hard wood	3,960	12,992
Glass	4,540	14,895
Steel	5,050	16,568

◀ These drums are made of steel.

This guitar is made of wood. ▶

1. How fast does sound travel in air? ____________

2. Does sound travel faster in glass or wood? ____________

3. Which carries sound faster in a guitar, the wood body or the steel strings? ____________

____

Discuss

What is your favorite musical instrument? What is it made of? What kinds of sounds does it make?

For more practice, go to pages 117–118.

Practice Pages

Name ______________________ Date ____________

Before You Read

VOCABULARY

A. Match each key word with its definition. Write the letter.

_____	**1.** echo	**a.**	how fast or slow the air vibrates
_____	**2.** pitch	**b.**	sound heard again
_____	**3.** sound waves	**c.**	how soft or loud a sound is
_____	**4.** frequency	**d.**	movements up and down
_____	**5.** volume	**e.**	vibrations that move through the air
_____	**6.** vibrations	**f.**	how high or low a sound is

B. Write six sentences using each key word and its definition.

1. ______________________

2. ______________________

3. ______________________

4. ______________________

5. ______________________

6. ______________________

C. Read the clues. Write the correct key words.

1. I am vibrations that travel through the air. What am I?

____

2. I am the loudness or softness of sound. What am I?

____

3. I am a sound that you hear again and again. What am I?

____

4. I am how low or high a sound is. What am I?

____

5. I am how fast or slow the air vibrates. What am I?

____

Science Reading Strategy: Act It Out

You learned that sound travels in waves. Read the paragraph. Act it out with several classmates.

If you throw a rock into the water, you will see waves moving away from the center. The waves near the center are closer together. As they move away from the center, the waves get farther apart. The same thing happens with sound. The closer you are to the source of a sound, the louder the sound will be. The farther away you are from the source, the softer the sound will be. Stand outside with your classmates. Have your classmates placed around you at different distances and shout loudly. Who can hear you the best? Who can't hear you as well?

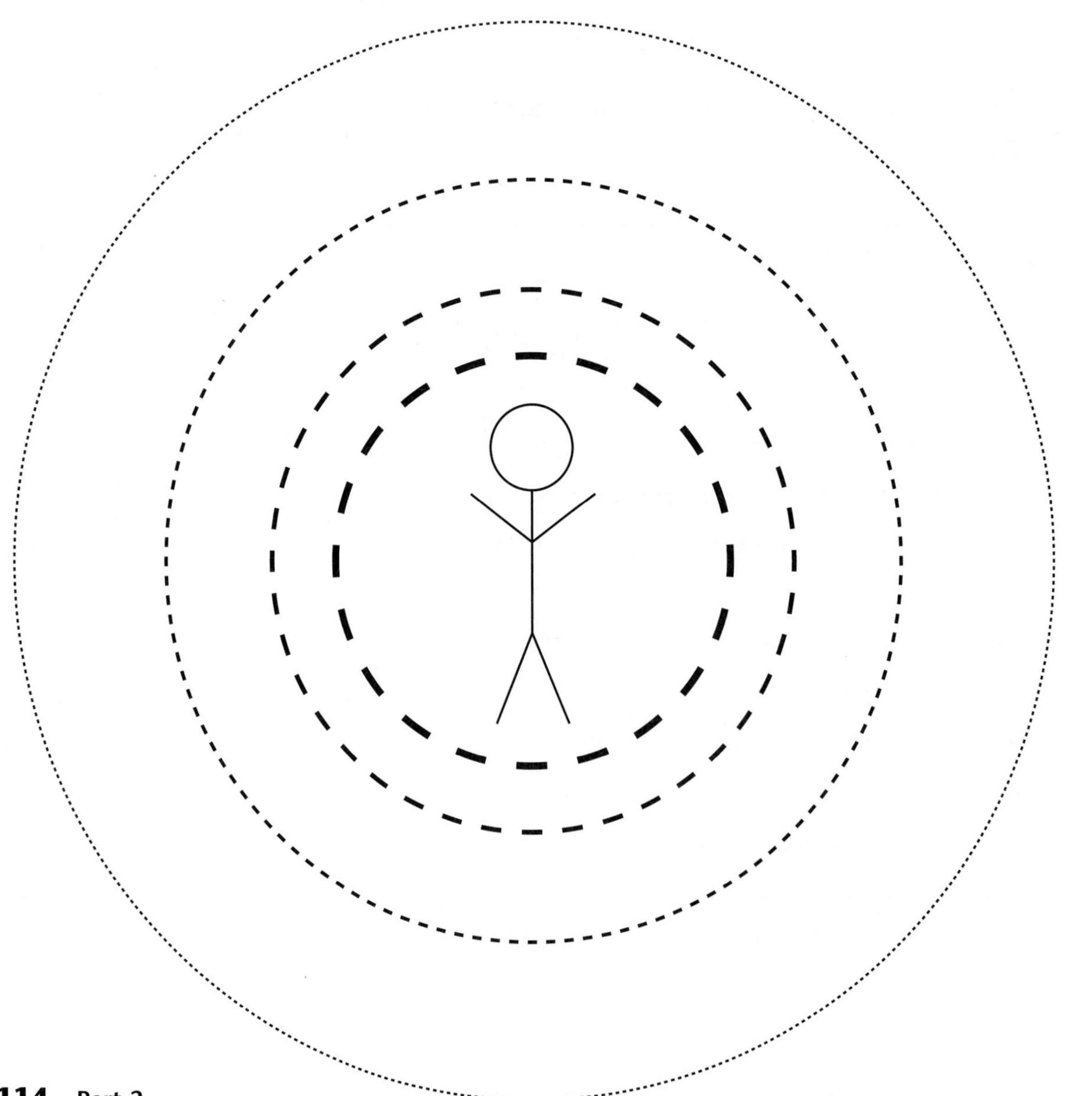

Name ____________________ Date __________

Using Visuals: Charts

A. Look at the charts. Compare and contrast the volume of various sounds. Then answer the questions.

Sound	Volume in Decibels
Plane taking off	150
Motorcycle	100
TV	70
Rain	50
Mosquito	20
Breathing	10

Sound	Volume in Decibels
Shuttle taking off	190
Rock concert	115
Busy street	70
Classroom or office	45
Falling leaves	10

1. Which sounds have the same volume in decibels?

   ____

2. Which sound has the highest volume?

   ____

3. Which sound is louder, falling leaves or rain?

   ____

4. How many decibels louder is a rock concert than a TV?

   ____

5. What decibel levels do you think are dangerous to your hearing?

   ____

PRACTICE

B. Look at the chart. Compare and contrast the hearing ranges of humans and other animals. Then answer the questions.

Species	Approximate Frequency Range of Hearing
Human	20–23,000 Hz
Bat	2,000–110,000 Hz
Beluga whale	1,000–123,000 Hz
Chicken	125–2,000 Hz
Dog	67–45,000 Hz
Dolphin	75–150,000 Hz

Hz=hertz (a unit of frequency)

1. Which animal's range of hearing is closest to a human's?

____

2. Which animal can hear sounds of the lowest frequency?

____

3. Which animal has the widest range of hearing?

____

4. Of the bat and the beluga whale, which animal has a wider range of hearing?

____

5. How does the chicken's hearing range compare to that of the other animals on the chart?

____

PRACTICE

Name ______________________________ Date ______________

Lesson 1 Review

VOCABULARY

Complete the puzzle. Use key words. Write the secret word.

1. ______ is measured in decibels. ___ ___ ___ ___ (___) ___
2. A whistle makes a high ______ wave. ___ ___ ___ ___ (___) ___ ___ ___ ___
3. ______ move out in all directions. (___) ___ ___ ___ ___ ___ ___ ___ ___ ___
4. ______ make sounds. ___ (___) ___ ___ ___ ___ ___ ___ ___ ___
5. An ______ is a repeated sound. ___ (___) ___ ___

Secret word: ___ ___ ___ ___ ___

VOCABULARY IN CONTEXT

Complete the paragraph. Use words from the box. There is one extra word.

pitch	echo	volume	vibrations	frequency

(1) ________________ is how high or low a sound is. It is caused by the **(2)** ________________ of the sound waves. A tuba makes low-frequency **(3)** ________________, so the sound it creates has a low pitch. Low sounds can be loud or soft. When a tuba player blows a lot of air into the horn, he or she makes a sound with a loud **(4)** ________________.

CHECK YOUR UNDERSTANDING

Choose the best answer. Circle the letter.

1. Pitch is caused by the ______ of the sound waves.
   a. frequency **b.** volume **c.** echo
2. Sound waves ______ out in all directions.
   a. throw **b.** dance **c.** move
3. Soft objects like carpets ______ sound waves.
   a. echo **b.** repeat **c.** absorb
4. Sound waves that are close together make a ______ sound.
   a. low-pitch **b.** high-pitch **c.** low-volume

PRACTICE

APPLY SCIENCE SKILLS

Science Reading Strategy: Act It Out

Read the paragraph. Act it out. Then answer the questions.

An echo occurs when a sound hits a hard surface and the waves bounce back. Use a small ball to act out this statement. Try throwing the ball against different types of hard surfaces.

1. Is there always an echo against any hard surface?

____

2. What would happen if the ball hit a pillow? Would there be an echo?

____

Using Visuals: Charts

Read the chart. Then answer the questions.

Matter	Speed of Sound	
	Meters per second	Feet per second
Dry, cold air	343	1,125
Water	1,550	5,085
Hard wood	3,960	12,992
Glass	4,540	14,895
Steel	5,050	16,568

1. In which kind of matter does sound travel the fastest?

____

2. In which kind of matter does sound travel the slowest?

____

3. Which kinds of matter are solids?

____

4. Does sound travel faster in water or in glass? Why?

____

5. Why does sound travel faster in steel than in dry, cold air?

____

PRACTICE

Science Journal

Write about five interesting things you have learned in this lesson.

1. ______________________________

____

____

2. ______________________________

____

____

3. ______________________________

____

____

4. ______________________________

____

____

5. ______________________________

____

____

Lesson 2 –Before You Read

Vocabulary

◀ This glass jar and the water in it are **transparent**. We can see through transparent things. Light can travel through them. Light is a type of **electromagnetic wave**. Electromagnetic waves have their own energy.

When light hits transparent things at an angle, it bends. The bending of light is called **refraction**. ▼

▲ This liquid is **translucent**. Light can travel through it. But we can't see through it clearly.

Key Words

- electromagnetic wave
- opaque
- reflection
- refraction
- translucent
- transparent

The moon is **opaque**. We can't see through an opaque object. Light can't travel through it. But some light waves bounce off its surface. This is called **reflection**. We can see the moon because it reflects the sun's light. ▼

LEARN MORE ABOUT IT

The light you see is one type of electromagnetic energy. Electromagnetic energy is a form of energy that travels through space in waves. The source of these waves is vibrating electric charges. These waves do not require a medium, so they can travel through a vacuum, or empty space. This is why you can see the sun and stars.

Practice

Next to each key word, write its meaning. Use your own words.

electromagnetic wave *the kind of wave that makes up light*

opaque ______________________

reflection ______________________

refraction ______________________

translucent ______________________

transparent ______________________

For more practice, go to page 135. 

Science Skills

Science Reading Strategy: **Draw a Picture**

You can **draw a picture** to help you understand what you read.

Read the text below. Look at the drawing of the first underlined sentence. Then draw a picture of the second underlined sentence.

When light hits a surface, some light waves bounce off. They bounce differently off different surfaces. Some surfaces, like mirrors, are very smooth. When light waves hit a smooth surface at an angle, they bounce off at the same angle.

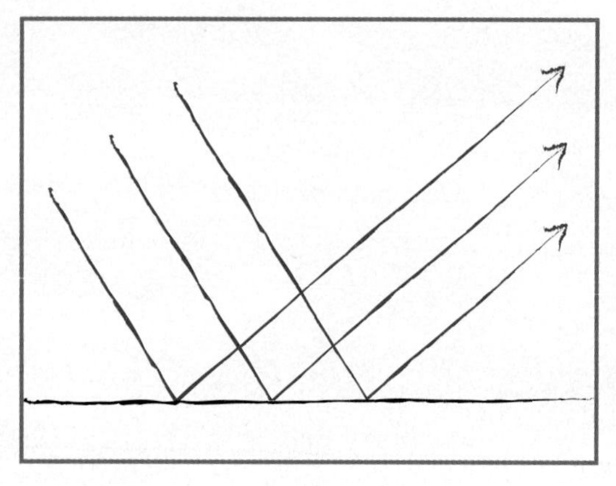

But most surfaces are not smooth. Your shirt, your pants, and your notebook paper have uneven surfaces. When light waves hit an uneven surface at an angle, they bounce off at all different angles.

Look for an **As You Read** activity in the lesson. It will ask you to draw a picture as you read. Your drawing will help you understand the lesson.

For more practice, go to pages 136–137. 

Using Visuals: **Wave Diagrams**

The first diagram below shows wavelength. The second diagram shows the colors of light we can see with our eyes, or visible light. Each color of light has a different wavelength and frequency.

Look at the diagrams. Then discuss the questions with a partner.

Wavelength is the distance between the top of one wave to the top of another. ▶

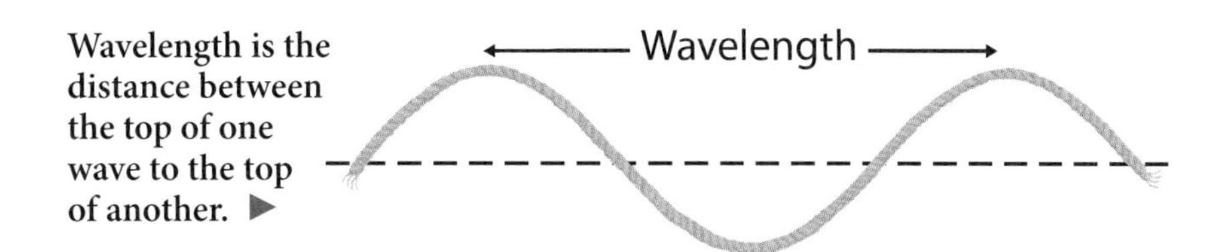

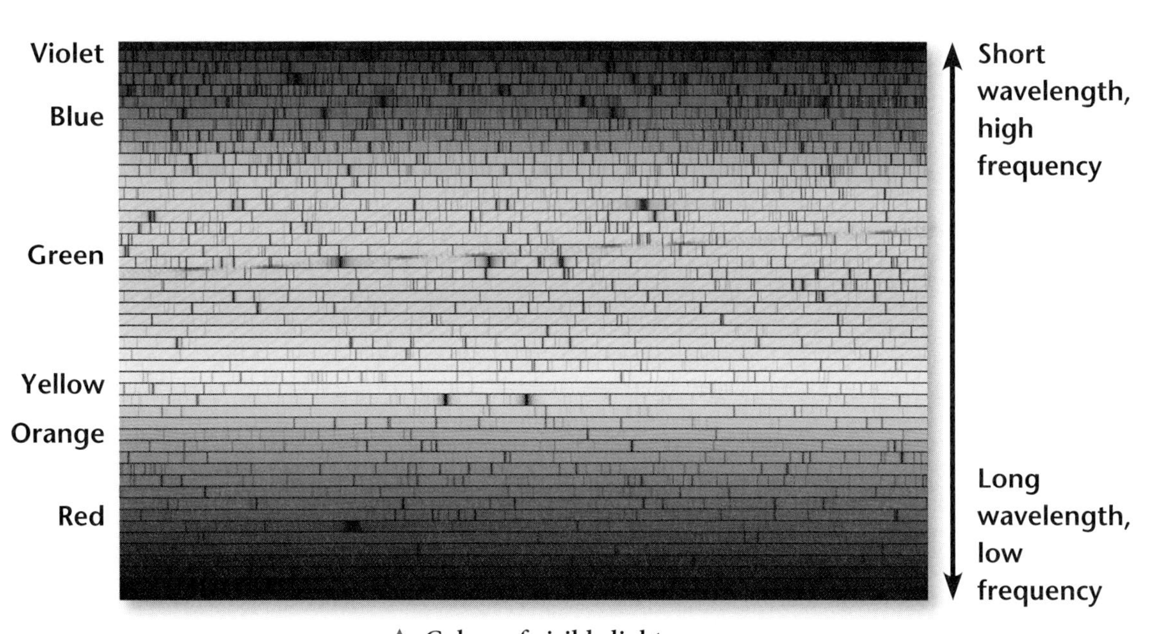

▲ Colors of visible light

1. Which color has the shortest wavelength?
2. Which color has the longest wavelength?
3. What frequencies do short and long wavelengths have?

For more practice, go to page 138.

Lesson 2

Light

Electromagnetic waves are all around us. Most electromagnetic waves are not **visible** to us. For example, your radio and TV use electromagnetic waves. Your toaster uses electromagnetic waves called infrared **rays** to toast your bread. Ultraviolet rays in sunlight can burn your skin. When your doctor wants to see your bones, he or she uses X rays. We can't see any of these waves.

Light is also an electromagnetic wave. But we can see light. It is a visible electromagnetic wave.

LANGUAGE TIP

you → your
he → his
she → her
we → our
they → their

---

visible: can be seen
rays: waves

Electromagnetic Spectrum

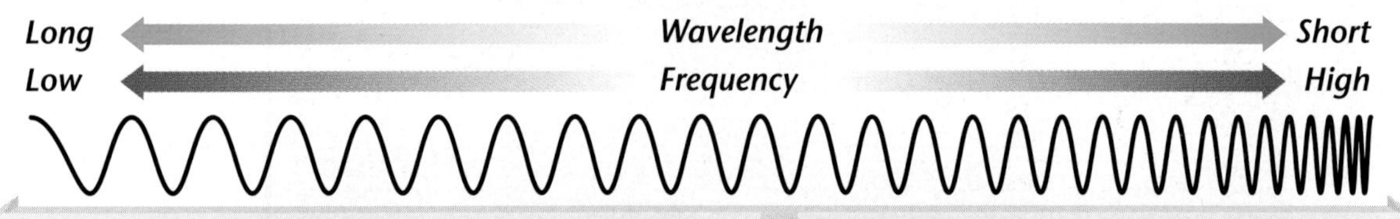

▲ Light is the only part of the electromagnetic spectrum that is visible.

Electromagnetic waves are different from sound waves. They have their own electrical and magnetic energy. And they don't need matter to travel through. They can travel through space. That's why we can see the sun, the moon, and other objects in space.

Electromagnetic waves travel very fast. They all travel at 300,000 meters (186,000 mi) per second. But the waves have different frequencies and wavelengths. We can't see most electromagnetic waves because their frequencies and wavelengths are too low or too high. But we can see the frequencies and wavelengths of light.

UV Rays

Ultraviolet rays, or UV rays, come from the sun. They are dangerous to living things in high amounts. UV rays can hurt our skin and eyes.

One of the layers of Earth's atmosphere protects us from UV rays. But pollution is causing this layer to become thin. In some places there are holes in this layer. Now more UV rays are reaching Earth's surface.

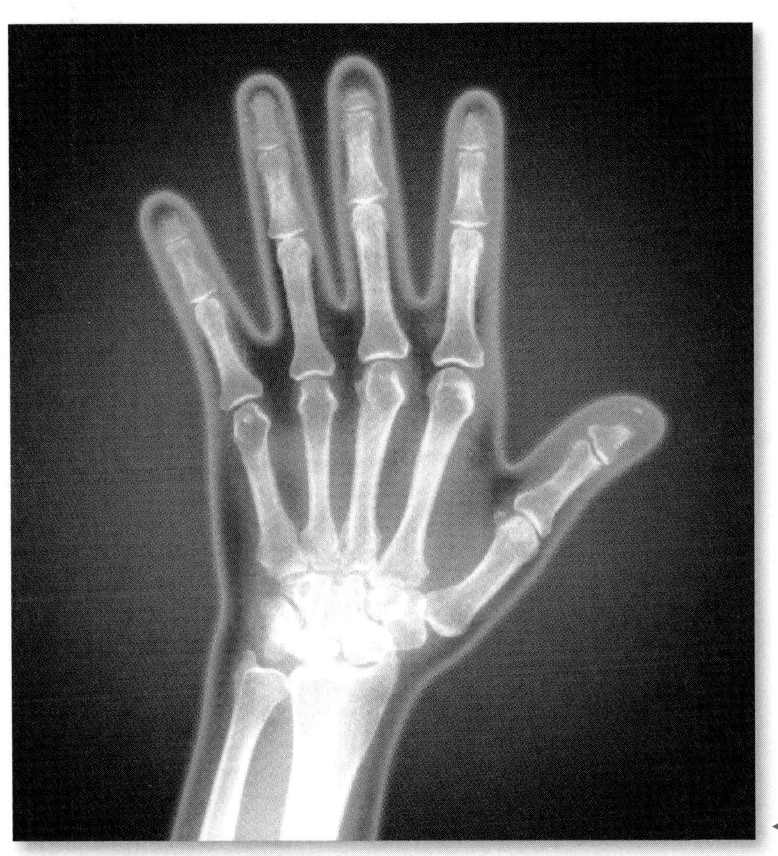

◀ Colored X ray of a hand

Before You Go On

1. What are two ways that electromagnetic waves are different from sound waves?
2. Which electromagnetic waves can we see?
3. Which electromagnetic waves can be dangerous?

Reflection

Light waves bounce off surfaces. This is called reflection. Reflection lets us see most objects. Light hits an object. Some of the light reflects off the object and reaches our eyes. We see the object.

When light hits a smooth surface, like a mirror, the light waves all bend together at the same angle. You see yourself because the light waves from your face bounce off the mirror together, as a unit.

When light waves hit an uneven surface, like your shirt, they bounce off at all different angles. This means the light waves go in all different directions. You can't see your reflection on an uneven surface.

SCIENCE AT HOME

Mirrors

We use many different mirrors every day. There are mirrors in our homes, in cars and buses, in our school, and in stores. Look around as you go about your day. How many mirrors do you see?

◀ **Light is bouncing off the kitten's face and hitting the glass. Then the light reflects off the glass. We see a reflection of the kitten's face.**

Light from candles is reflecting off the objects in this room. ▶

Transparency

Glass and clean water are transparent materials. Light waves can pass through them. We can see other objects through transparent materials.

Wood and metal are opaque materials. Light waves cannot travel through them. We can't see through opaque materials. They **absorb**, or take in, some light and reflect the rest of it.

Some materials aren't opaque or transparent. They are translucent. Frosted glass is a translucent material. Light waves can pass through it, but they break up and move in different directions. That's why you can't see all the way through a translucent material.

---

absorb: to take in, as a sponge takes in water

LANGUAGE TIP

A *material* is a kind of matter. We group together materials by properties they share.

▲ This frosted glass is translucent.

◀ This glass is transparent.

Before You Go On

1. What allows you to see most objects?
2. What are some examples of transparent, opaque, and translucent materials?
3. How is a mirror different from plain glass?

Colors of Light

There are different colors of light. The color of light **depends on** its wavelength and frequency. Red light has the longest wavelength and the lowest frequency. Violet light has the shortest wavelength and the highest frequency. White light is all the different wavelengths of visible light together.

LANGUAGE TIP

wave + length = wavelength

The words *wave* and *length* combine to make a new word.

---

depends on: is caused by; is because of

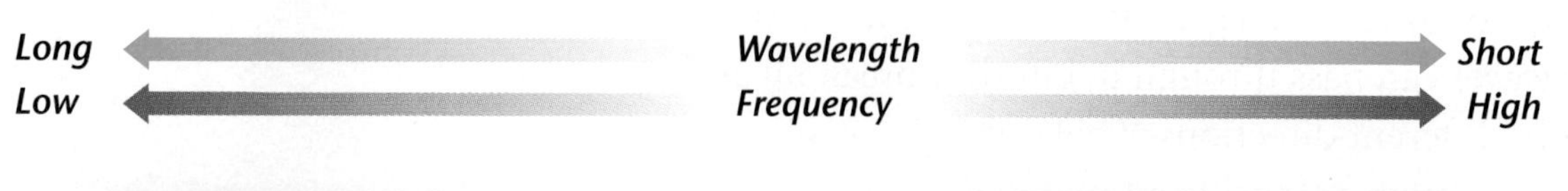

▲ This is a prism. When white light passes through it, the light bends and then breaks, or refracts, into different colors.

Colors of Objects

An object's color depends on light. Each object absorbs certain wavelengths of light. The object reflects the other wavelengths. The reflected light gives the object its color.

When white light hits a red apple, the apple absorbs all the other wavelengths except the red. The apple reflects the red wavelength.

In the dark, there is no visible light. Objects do not appear to have colors.

◀ **Different feathers on this bird reflect different wavelengths of light.**

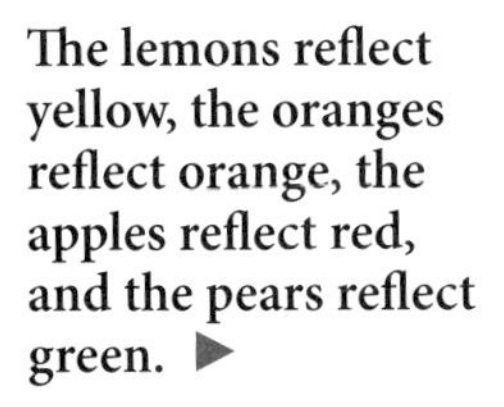

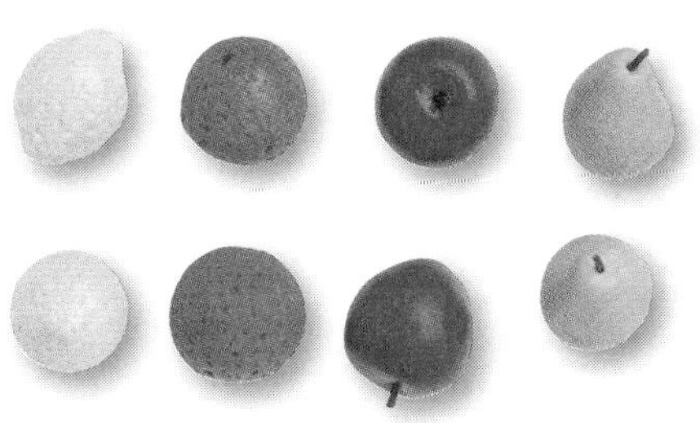

The lemons reflect yellow, the oranges reflect orange, the apples reflect red, and the pears reflect green. ▶

As You Read

Draw a Picture

Draw a picture of a red apple absorbing and reflecting the different wavelengths of light.

SCIENCE NOW

Lasers

A laser makes a very focused stream of light. The light waves travel in only one direction. Machines we see every day use lasers. CD players use lasers to read the tiny bumps on a CD. And a supermarket clerk uses lasers to read the prices of items at the checkout counter.

▲ **Laser beam reading a CD**

Before You Go On

1. What is white light?
2. What are the names of the different colors of light?
3. What wavelengths does a lemon absorb?

Refraction

Refraction is the bending of light. Light refracts, or bends, when it goes from one type of matter into another at an angle. The light bends because it slows down or speeds up in each type of matter.

Here is an example. Light traveling through the air hits the water in an aquarium at an angle. The light slows down in the water and refracts. The light refracts again in the aquarium's glass walls. Then it refracts again as it speeds up in the air outside the aquarium.

▲ This light is refracting in an aquarium.

HISTORY CONNECTION

Telescopes and Galileo

Scientists use telescopes to study objects in space. Telescopes use glass to reflect and refract light. This can help you see faraway objects. From 1609 to 1610, Galileo used a telescope to study the moon and the planets. He learned that Earth moves around the sun.

▲ Galileo

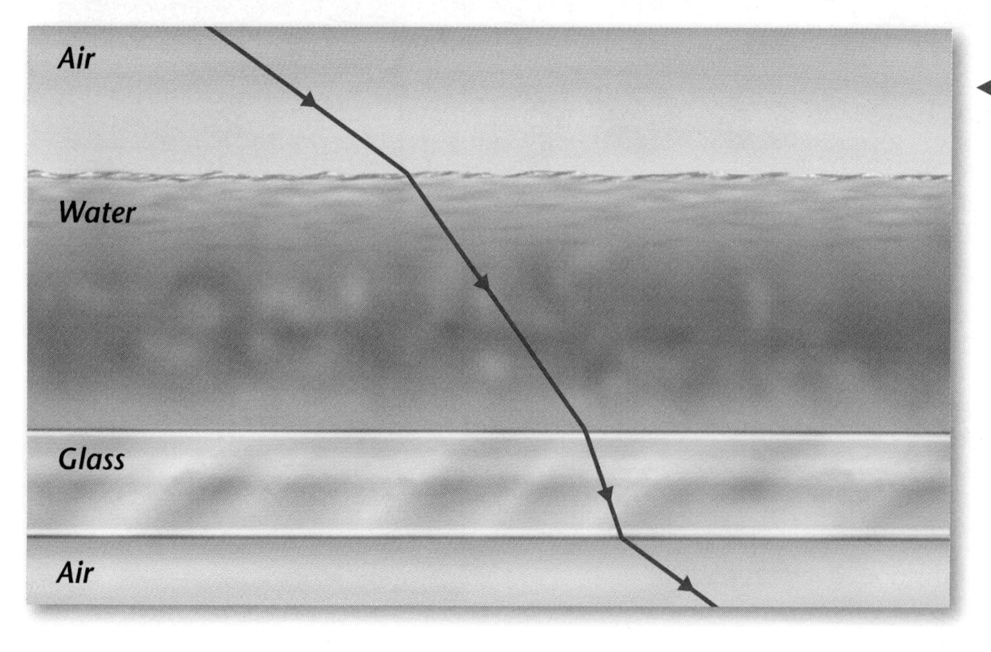

◀ This diagram shows light refracting in the different types of matter in an aquarium.

Prisms and Rainbows

How does a prism change white light into the colors that make it up? It refracts each wavelength of light at a different angle, from the longest wavelength to the shortest.

Water also acts like a prism. Rainbows are made when tiny drops of water in the air separate the sunlight into its different wavelengths.

We can see a rainbow when tiny drops of water in the air refract sunlight. ▶

Leaders in Science

Thomas Edison (1847–1931)

Thomas Edison invented the modern lightbulb. Before lightbulbs, people used candles or gas to light their homes. In 1879, Edison made a lightbulb that lasted 14 hours. In 1880, he invented a bulb that lasted 1,200 hours. Two years later he opened the first modern electrical power station. The station was in New York City and had fifty-nine customers.

▲ Edison touches an early lightbulb. He holds a smaller, better bulb.

Before You Go On

1. When does light refract?
2. How does a prism work?
3. Why are the colors in a rainbow always in the same order?

Lesson 2 – Review and Practice

Vocabulary

Complete each sentence with the correct word or phrase from the box.

reflection	opaque	transparent
translucent	refraction	electromagnetic wave

1. The bending of light is called ________________.
2. Light bounces off a surface. This is called ________________.
3. Light waves can pass through ________________ materials.
4. Frosted glass is an example of a ________________ material.
5. Some light is absorbed and some light reflects off ________________ materials.
6. Light is a type of ________________.

Check Your Understanding

Write the answer to each question.

1. What's the only electromagnetic wave we can see? ________________
2. What does the color of light depend on? ________________

____

3. What wavelengths does a white shirt reflect? ________________

Apply Science Skills

Science Reading Strategy: **Draw a Picture**

Reread the definition of *translucent* on page 127. Draw a picture of light waves traveling through the translucent liquid shown on page 120.

Using Visuals: **Wave Diagrams**

These diagrams show light waves hitting two different types of surfaces. Look at the diagrams. Then answer the questions.

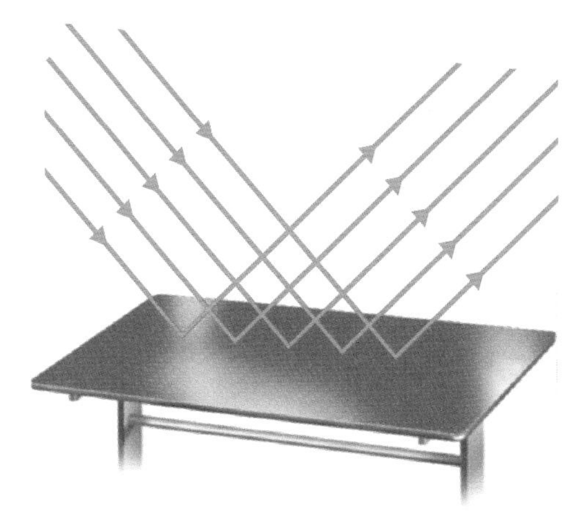

▲ Light hitting a smooth surface

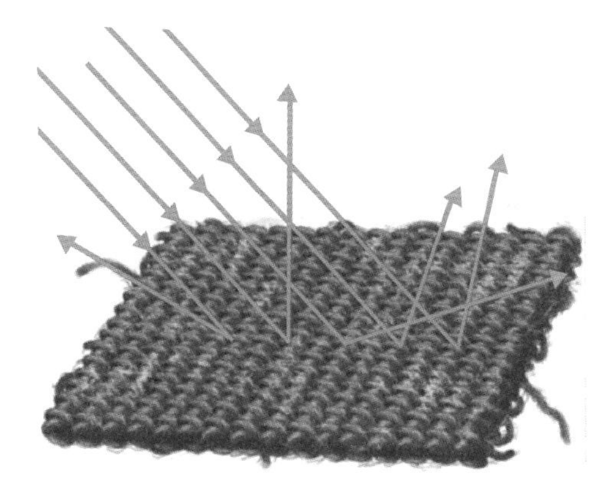

▲ Light hitting an uneven surface

1. In what type of surface can you see a reflection of yourself? ______________________________

____

2. What happens when light waves hit an uneven surface? ______________________________

____

3. What are some examples of smooth and uneven surfaces? ______________________________

____

Discuss

Red, orange, and yellow are called warm colors. They feel warm and exciting. Green, blue, and violet are called cool colors. They feel cool and relaxing. Which group do you like more? Why?

For more practice, go to pages 139–140.

Practice Pages

PRACTICE

Name ______________________ Date __________

Before You Read

VOCABULARY

A. Match each key word with its definition. Write the letter.

_____ **1.** reflection	**a.** the bending of light
_____ **2.** refraction	**b.** the bouncing of light waves off a surface
_____ **3.** transparent	**c.** light is one type
_____ **4.** translucent	**d.** you can see through it
_____ **5.** opaque	**e.** you cannot see through it clearly
_____ **6.** electromagnetic wave	**f.** light cannot pass through it

B. Write six sentences using each key word and its definition.

1. ______________________

2. ______________________

3. ______________________

4. ______________________

5. ______________________

6. ______________________

C. Write T for *true* or F for *false.* Correct the sentences that are false.

_____ **1.** Transparent objects reflect light. ______________________

_____ **2.** Electromagnetic waves have their own energy. ______________________

____

_____ **3.** The moon is a translucent object. ______________________

_____ **4.** We can see through transparent objects. ______________________

_____ **5.** Opaque means that light cannot pass through it. ______________________

____

_____ **6.** Light is a type of sound wave. ______________________

PRACTICE

Science Reading Strategy: Draw a Picture

A. Read the definition of *reflection*. Draw a picture below to represent it.

Light waves bounce off the surface of opaque objects like the moon. We can see the moon because it reflects the sun's light. This is called reflection.

B. Show your picture to a partner. Talk about your pictures.

Name ______________________ Date ____________

C. Read the paragraph. Draw a picture below of the underlined sentences.

Electromagnetic waves are all around you. Light is a kind of electromagnetic wave. When you go to the beach, the sun's ultraviolet (UV) rays can burn your skin. <u>UV rays come to you in waves and hit your skin repeatedly. You cannot see these waves, but you can feel the heat. If you stay too long under the sun, these light waves will harm you.</u>

D. Show your picture to a partner. Talk about your pictures.

Using Visuals: Wave Diagrams

Look at the wave diagram of the electromagnetic spectrum. Then answer the questions.

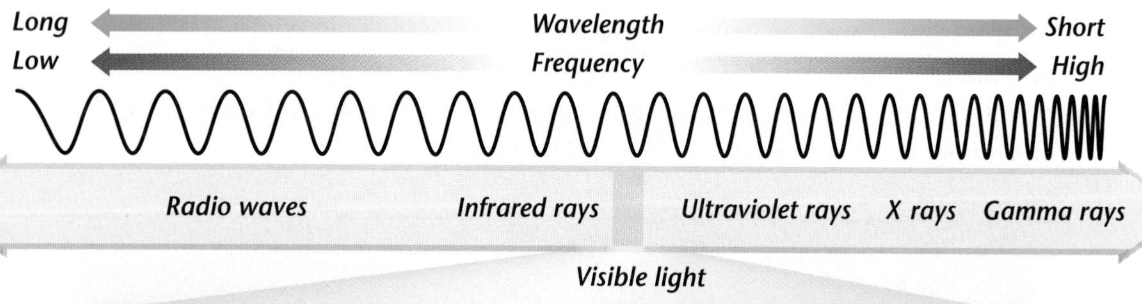

1. List the different types of waves on the diagram.

____

____

2. Which electromagnetic waves have the highest frequency?

____

3. Which waves have the longest wavelength?

____

4. Do infrared rays have a short or long wavelength?

____

5. Do you think that high-frequency waves can be harmful? Why or why not?

____

____

Name ______________________ Date __________

Lesson 2 Review

VOCABULARY

Use words from the box to identify each clue. Write the word on the line.

reflection	refraction	opaque	electromagnetic waves	transparent

1. I am the bending of light. ______________
2. I don't let light go through me. I am... ______________
3. I am visible light, X rays, and radio waves. ______________
4. You can see through me. I am... ______________
5. I am light waves bouncing off surfaces. ______________

VOCABULARY IN CONTEXT

Circle the correct word or phrase to complete each sentence.

Light is (**1.** a sound / an electromagnetic) wave. Water drops can make a rainbow by (**2.** reflecting / refracting) light. The drops of water separate the sunlight into different wavelengths. But water can also cause a (**3.** reflection / refraction). Think of looking into smooth water on a lake. You can see yourself. When lake water is (**4.** opaque / transparent), you can see to the bottom. Some lake water might have mud in it. You cannot see through it clearly. It is (**5.** transparent / translucent).

CHECK YOUR UNDERSTANDING

Choose the best answer. Circle the letter.

1. Toasters use ______ rays to toast waffles.

a. opaque **b.** ultraviolet **c.** infrared

2. Electromagnetic waves have their own ______ and magnetic energy.

a. transparent **b.** electrical **c.** wavy

3. The color of light depends on its ______ and frequency.

a. refraction **b.** wavelength **c.** reflection

4. You can't see a reflection of yourself on ______ surface.

a. an uneven **b.** an even **c.** a hard

PRACTICE

APPLY SCIENCE SKILLS

Science Reading Strategy: Draw a Picture

Find an object that is mostly one color. Draw a picture of this object absorbing and reflecting the different wavelengths of light.

Using Visuals: Wave Diagrams

Look at the wave diagrams on page 133. Then answer the questions.

1. What happens when light waves hit a smooth surface?

   ____

2. Which wave diagram shows what happens when light waves hit a mirror's surface?

   ____

3. Which wave diagram shows what happens when light hits the surface of your clothing?

   ____

PRACTICE

Science Journal

Write about five interesting things you have learned in this lesson.

1. ______________________________

____

____

2. ______________________________

____

____

3. ______________________________

____

____

4. ______________________________

____

____

5. ______________________________

____

____

Part Review

Vocabulary

Answer the questions with words or phrases from the box.

frequency	reflection	opaque	sound waves
pitch	vibrations	volume	refraction
translucent	electromagnetic wave	echo	transparent

1. What makes sound waves? ______________________

2. How do vibrations travel through air? ______________________

3. What do sound waves bouncing off a hard surface make? ______________________

4. What is light bouncing off a surface called? ______________________

5. What is the bending of light called? ______________________

6. What kind of wave has electrical and magnetic energy? ______________________

7. What is the loudness or softness of sound called? ______________________

8. What is how high or low something sounds? ______________________

9. What causes pitch? ______________________

10. What kind of material can light go all the way through? ______________________

11. What kind of material absorbs some light and reflects the rest? ______________________

12. What kind of material is frosted glass? ______________________

Check Your Understanding

Write *T* for *true* and *F* for *false*. Then, with a partner, discuss how to make each false statement true.

_______ 1. Sound travels fastest through gases.

_______ 2. Something vibrating quickly makes high-frequency sound waves.

_______ 3. Sound waves need matter to travel, but light waves don't.

_______ 4. You can see yourself in a mirror because the light is refracted.

Extension Project

Shine a flashlight through a prism. Watch how white light is separated. Draw a picture of what you see.

Apply Science Skills

Using Visuals: **Wave Diagrams**

Look at the electromagnetic wave diagram. Write the numbers 1–6 in your notebook. Next to each number, write the correct label from the box.

visible light	radio waves	UV rays
X rays	gamma rays	infrared rays

1 2 3 4 5 6

▲ **Electromagnetic spectrum**

1. Which waves does your toaster use? _______________

2. Which waves does a doctor use to see your bones? _______________

3. Which waves can you see? _______________

Part Experiment

How Does Light Reflect and Refract?

Remember what you learned about the reflection and refraction of light. Light waves bounce off smooth surfaces all together, at the same angle. Light waves can bend when they hit different kinds of matter at an angle. The waves of white light can be separated into different wavelengths.

Purpose

To observe the reflection and refraction of light

Materials

flashlight
mirror
pencil
clear glass of water
CD

What to Do

1. Turn to page 146 for a copy of the data table. Then darken the room. Shine the flashlight on the mirror at different angles. Record your observations.
2. Half fill the glass with water. Place the pencil in the glass at an angle. Look at the glass from the side. Record your observations.
3. Hold up the glass of water 6–12 inches from your face. Look through the water at different objects in the room. Record your observations.
4. Hold the CD at different angles under light. Record your observations.

Data Table	
Step	**Observations**
1	
2	
3	
4	

Draw Conclusions

Work with a partner. Discuss your conclusions. Then write them under Step 5 on page 146.

1. How did the light reflect off the mirror? Why?
2. Why did the pencil look different in the water?
3. Why did objects look different through the glass of water?
4. What happened to light when it hit the CD? Why?

Experiment Log:

How Does Light Reflect and Refract?

Follow the steps of the scientific method as you do your experiment. Write notes about each step as the experiment progresses.

Step 1: Ask questions.

Step 2: Make a hypothesis.

Step 3: Test your hypothesis.

Step 4: Observe.

Data Table	
Step	**Observation**
1.	
2.	
3.	
4.	

Step 5: Draw conclusions.

Write About It

1. Write a hypothesis about the relationship between the different light colors and the colors of the coolest and hottest stars. Look at the electromagnetic spectrum diagram on page 124 or 128 to help you write your hypothesis.

2. In your opinion, which is more important—to be able to see objects or to hear sounds? Write a paragraph explaining why.

Glossary

Phonetic Respelling Key

a	cat	o	stop
ah	father	oh	go, slow, toe
air	hair, there, their	oo	moon, blue, do
ar	arm	yoo	you, music, few
ay	play, make, eight, they	or	for, your
aw	draw, all, walk	oi	soil, boy
e	red, said	ow	brown, out
ee	green, please, she	u	put, look
eer	ear, here	uh	but, what, from, about, seven
eye	like, right, fly	er	her, work, bird, fur
i	six		

absorb (ab-SORB)
To take in and hold, like a sponge takes in and holds water.

act it out (akt it OWT)
A reading strategy. Act out ideas in the text. For example, to understand how your vocal cords vibrate, make a sound and touch your throat. Then you can feel the vibrations your vocal cords make.

ask questions (ask KWES-chuhnz)
1. A reading strategy. Ask yourself questions about the text. This helps you check your understanding. **2.** Step 1 of the scientific method. The purpose of an experiment is to find out answers to the questions asked in step 1.

atoms (AT-uhmz)
The tiny pieces that make up all matter.

boiling point (BOI-ling point)
The temperature at which a solid changes to a gas. Water boils at 100° Celsius (212° Fahrenheit).

cause and effect (kawz and uh-FEKT)
A reading strategy. Looking for cause and effect can help you understand a text better. A cause makes something happen. The effect is what happens.

chart (chart)
A kind of visual. A chart often shows information in columns and rows.

chemical change (KEM-uh-kuhl chaynj)
A change in which the groups of atoms that make up something change. When metal rusts, it is a chemical change.

cycle diagram (SEYE-kuhl DEYE-uh-gram)
A kind of visual. A cycle diagram is a labeled drawing, often in the shape of a circle. It shows events that happen again and again in the same order.

decibels (DES-uh-buhlz)
Units used to measure volume.

density (DEN-suh-tee)
The amount of mass something has per unit of volume. A cup of rocks has a higher density than a cup of feathers. To figure out density, you divide mass by volume.

depends on (duh-PENZ on)
Is caused by; is because of.

dissolves (duh-ZOHLVZ)
Changes from solid to liquid form.

draw conclusions
(draw kuhn-KLOO-zhuhnz)
Step 5 of the scientific method. You draw conclusions at the end of an experiment. You decide whether your hypothesis is correct.

echo (EK-oh)
A sound that repeats. You can hear an echo in a place with hard surfaces. This happens because the sound waves bounce off the hard surfaces.

electromagnetic energy
(uh-LEK-troh-mag-NET-ik EN-uhr-jee)
A form of energy that travels through space in waves.

electromagnetic waves
(uh-LEK-troh-mag-NET-ik wayvz)
The kind of waves that make up light. Light waves are the only electromagnetic waves we can see.

facts and examples
(fakts and ig-ZAM-puhlz)
A reading strategy. As you read, look for facts and examples. A fact is a true statement about a topic. An example is a statement that helps illustrate or explain a fact.

frequency (FREE-kwuhn-see)
How fast or slow a sound wave moves. A high-frequency sound wave moves fast. A low-frequency sound wave moves slow.

gas (gas)
One of three different states of matter. A gas does not have a set volume or a set shape. A gas gets bigger in order to fill its container.

idea maps (eye-DEE-uh maps)
A reading strategy. You make an idea map to show how ideas are connected. You list the main ideas first. You connect these ideas to details and examples.

illustration (il-uhs-TRAY-shuhn)
A kind of visual. An illustration is a drawing or a painting.

liquid (LIK-wid)
One of three different states of matter. A liquid has a set volume but not a set shape.

make a hypothesis
(mayk uh heye-PAH-thuh-suhs)
Step 2 of the scientific method. A hypothesis is a guess. You guess what the experiment will show. Then you do the experiment to see if your hypothesis is correct.

mass (mas)
The amount of something there is. Mass can be measured with a balance. Mass is measured in grams and kilograms.

matter (MAT-uhr)
Anything that takes up space and has mass. Water, oxygen, salt, and sand are examples of different types of matter. Matter can be a pure substance or a mixture of different substances.

measure (MEZH-uhr)
To find out facts about something by using standard tools. Different tools measure different things. For example, a ruler measures length, or how long something is. A balance measures mass. Special containers measure volume.

melting point (MEL-ting point)
The temperature at which a solid changes to a liquid. Ice melts at 0° Celsius (32° Fahrenheit).

micrograph (MEYE-kroh-graf)
A kind of visual. A micrograph is a photograph taken with the help of a microscope. We use microscopes to see small things up close.

observe (uhb-ZERV)
Step 4 of the scientific method. To observe is to watch something in order to learn about it. When you do an experiment, you write down what you observe.

opaque (oh-PAYK)
Opaque materials do not let light go through. You can't see through them. Wood and milk are opaque materials.

particle (PAHR-ti-kuhl)
A very small piece of matter.

photo sequence (FOH-to SEE-kwuhns)
A kind of visual. A photo sequence is a set of two or more photos. It shows stages of an event that happens over time.

physical change (FIZ-uh-kuhl chaynj)
A change in which matter changes in form only. The matter itself does not change. An example is water boiling. Water changes from a liquid to a gas, but it is still water.

pie charts (peye charts)
A kind of visual. A pie chart looks like a pie with cut slices. The slices add up to 100 percent (%). Each slice shows part of this 100 percent.

pitch (pich)
How high or low a sound is. Mice make squeaky, high-pitched sounds. Elephants make deep, low-pitched sounds.

predict (pruh-DIKT)
A reading strategy. Before you read, you can predict, or guess, what the text is about. This will help focus your attention as you read the text.

preview (PREE-vyoo)
A reading strategy. To preview a text, look at the headings, pictures, and captions before you start reading. This will help focus your attention.

properties (PROP-uhr-teez)
Characteristic traits or qualities. For example, atoms look and act in particular ways depending on what state of matter they are in. The atoms in liquids look and act differently from atoms in solids.

rays (rayz)
Types of waves.

reflection (ruh-FLEK-shuhn)
The bouncing of light off an object. Reflection allows us to see objects.

refraction (ruh-FRAK-shuhn)
The bending of light as it passes through transparent matter. Water, glass, and prisms can all refract light.

reread (ree-reed)
A reading strategy. To reread is to read again. Reread difficult parts of the text to understand them better.

similar (SIM-uh-ler)
Having a likeness or resemblance; alike.

solid (SOL-id)
One of three different states of matter. A solid has a set shape and volume.

sound waves (sownd wayvz)
Vibrations moving through matter, including air. When a sound wave reaches our ears, we hear a sound.

states (stayts)
Forms. The three states of matter are solid, liquid, and gas.

states of matter illustration (stayts uhv MAT-uhr il-uh-STRAY-shuhn)
A kind of visual. These illustrations show how atoms look and act in solids, liquids, and gases.

substance (SUB-stuhns)
Physical material or matter.

test the hypothesis
(test thuh heye-PAH-thuh-suhs)
Step 3 of the scientific method. You test the hypothesis to see if it is correct. Doing an experiment is how a scientist tests his or her hypothesis.

tools (toolz)
What scientists use to observe and measure things. A thermometer, a ruler, and a microscope are all tools.

translucent (tranz-LOO-suhnt)
A translucent material lets some light go through. We can't see through translucent materials clearly. Frosted glass is translucent.

transparent (tranz-PAIR-uhnt)
Clear. Light can travel through transparent matter. You can see through transparent materials such as glass, water, and air.

ultraviolet rays (Ul-tra-vi-o-let rayz)
Rays that come from the sun. They are dangerous to living things in high amounts. Also called UV rays.

vibrations (veye-BRAY-shuhnz)
Up-and-down movements. Drums make vibrations. Vibrations move through matter, making it vibrate, too.

visible (VIZ-uh-buhl)
Able to be seen.

visualize (VIZH-yoo-uh-leyez)
A reading strategy. To visualize is to picture things in your mind. As you read, visualize the ideas in the text.

volume (VOL-yoom)
1. **(matter)** The amount of space something takes up. You use special containers to measure volume.

2. **(sound)** The loudness or softness of a sound. Sounds at a low volume are soft. Sounds at a high volume are loud.

wave diagram (wayv DEYE-uh-gram)
A kind of visual. Wave diagrams are labeled drawings of the waves that make up sound or light. Wave diagrams often show wavelength and frequency.

Index

M

O

P

R

S

T

U

V

W

X

Credits

Illustrations

Page 6, 23 (top right), 40 (top left), 103, 109, 115, 118, 120, 124, 126 (top/bottom), Shutterstock Images; 22 (top right), Simone End/Dorling Kindersley; 22 (bottom left), Laurie O'Keefe/Pearson Education/Prentice Hall College Division; 22 (bottom right), Peter Bull/Dorling Kindersley; 24, 25, Michelle Ross/Dorling Kindersley; 39 (top left), Walter Stuart; 40 (bottom right), Tom Leonard; 75 (left top, middle, bottom), 88 (bottom), Ricky Blakeley/Dorling Kindersley; 77 (bottom), 106 (bottom), 108 (bottom), Dorling Kindersley; 78, North Wind Picture Archives; 86, Morgan Cain & Assocs.; 91, Richard Lewis/Dorling Kindersley; 98 (top right), David A. Hardy, PhotoResearchers, Inc.; 102, John Edwards & Assocs.; 103, Network Graphics/Pearson Education/Prentice Hall College Division; 123 (top), Morgan Cain & Assocs.; 123 (bottom), National Oceanic and Atmospheric Administration; 124, Martucci Design; 130 (top), W.J. Short/EMG Education Management Group; 130 (bottom), 133, Precision Graphics; 131 (right center), Bettman CORBIS; 143, Janos Marffy/Dorling Kindersley.

Photographs

All photos are from Shutterstock.com except as otherwise noted.

COVER (background) Shutterstock.com, (top left) Shutterstock.com, (top right) Shutterstock.com, (bottom left) Shutterstock.com, (bottom right) Shutterstock.com.

GETTING STARTED Page 2 (left), 4 (right bottom), 9 (middle right), 31 (middle bottom), NASA; 6 (top), Photo Researchers/Alamy; 10 (left/right), 11 (left/right), Steve Gorton and Gary Ombler/Dorling Kindersley; 14 (top/bottom), 35, Steve Cole/Photodisc/Getty Images; 14 (middle top), 35, Mike Dunning/Dorling Kindersley; 14 (middle bottom), 35, Andy Crawford/Dorling Kindersley; 15 (top), 35, Ryan McVay/Photodisc/Getty Images; 15 (middle top), 35, Photodisc/Getty Images; 15 (middle bottom), 35, ©1991 Paul Silverman/Fundamental Photographs, NYC; 15 (bottom), 35, Pearson Learning Photo Studio; 25 (left), Prentice Hall School Division; 25 (bottom right), Pearson Education Digital Archive; 27 (top left), Runk Schoenberger/Grant Heilman Photography, Inc.; 27 (middle left), Dorling Kindersley; 27 (middle right), Matthew Ward/Dorling Kindersley; 27 (right), Andrew Butler/Dorling Kindersley; 31 (top right), Monique le Luhandre/Dorling Kindersley.

PART 1 Page 44 (bottom left), 55, Leslie Weidenman; 44 (bottom right), Doug Scott/age footstock america; 45, 52 (top right), Prentice Hall School Division; 46 (bottom left), Dave King/Dorling Kindersley; 47, Naval Research Laboratory/US Naval Historical Center (CuP) Photography; 49 (top left), Richard Walters/Visuals Unlimited; 49 (top right), Marli Miller; 49 (bottom left), Andrew Syred/Allstock/Getty Images; 49 (bottom right), Dreamstime; 52 (bottom), Pearson Learning Photo Station; 53 (right), 73 (left), 74 (bottom right), 92 (bottom left), 92 (middle left), 92 (bottom right), 93 (left/right), Clive Streeter/Dorling Kindersley; 54 (top left), Alan Hills and Barbara Winter/Dorling Kindersley/©The British Museum; 54 (top right), Erich Lessin/Art Resource; 54 (bottom right), Jean-Loup Carmet/Photo Researchers, Inc.; 57 (top left), Georges Seurat/The Art Institute of Chicago/Georges Seurat, A Sunday Afternoon on the Island of La Grande Jatte. 1884-86. Oil on canvas. 6'9 ½"x10'1 ¼". Helen Birch Bartlett Memorial Collection. Photograph ©2005, The Art Institute of Chicago. All rights reserved; 57 (bottom left), Almaden Research Center/IBM Research (unauthorized use is prohibited); 59 (top right), Colin Cuthburt/Science Photo Library/Photo Researchers, Inc; 66 (top left), 77 (bottom), Dorling Kindersley; 73 (top right), Tony Freeman/PhotoEdit; 78 (left), Peter Chadwick/Dorling Kindersley; 78 (right center), ©Judith Miller/Dorling Kindersley; 79 (top right), Peter Gardner/Dorling Kindersley.

PART 2 Page 98 (top), 120 (bottom left), 144 (top right), Dorling Kindersley; 100 (bottom right), Prentice Hall School Division; 101 (top left), Pearson Education Corporate Digital Archive; 107 (top right), Jane Miller/Dorling Kindersley; 107 (bottom left), Irv Beckman/Dorling Kindersley; 120 (bottom right), Dreamstime; 129 (center), Alex Wilson/Dorling Kindersley; 130 (top left), ©1998 Richard Megna, Fundamental Photographs, NYC; 144 (top and bottom left), Tim Ridley/Dorling Kindersley; 144 (top middle), Peter Gardner/Dorling Kindersley; 144 (bottom right), Mike Dunning/Dorling Kindersley.